PRAISE FOR ONE MORE TIME:

"John Posey wrote *One More Time* for the man who reached the back half of life and quietly decided the fight was over. It tears down the lie that your best days are spent and God has moved on to someone younger. His word to every man who thinks he is finished: 'Do you have any scar tissue? Then you're qualified.' Read it. Then go pull someone out of the water. One more time."

- Christopher C. Simpson, President & CEO, CBMC International, US Secret Service (Ret.), US Marine Corps

"*One More Time* is an inspiration for all of us to pour into the next generation, and in the process, as Paul says to fight the good fight, to finish the course, and to remain faithful, until we go to be with Jesus."

- Boyd Bailey, President, NCF Georgia | Founder, Wisdom Hunters

"*One More Time* is a book about finishing well, discovering God's purpose in every season of life. John's reflections on Chronos and Kairos, and condemnation to righteousness are dynamic and practical. I recommend this book to followers of Jesus who want to hear 'Well done' when they finish their race."

- Kenneth Boa, Founder, Reflections Ministries and Museum of Created Beauty

"I wholeheartedly endorse your book because it is

much needed as a guide for those of us at this stage of our journey. More importantly, it helps those around us understand our challenges so they may encourage us to 'finish well.' Lastly, I know the book's truth, having been blessed to witness your ministry and mentorship. I pray that God will bless you as you share this work with many 'one more timers'!

- Charles W. Buffington, CEO and Founder, Sonseeker Media Partners

"John Posey's book offers a compelling reminder that every life has purpose, every scar can carry meaning, and it is never too late to finish well. With biblical encouragement, practical wisdom, and a message of hope, it inspires readers to trust God's plan and embrace the next chapter of their journey with faith and confidence."

- Dr. Terence Chatmon, Victorious Family

"John Posey reminds us that these years are not for retreat but for renewed purpose. I highly recommend *One More Time* to anyone who wants to finish well and discover that God may still have some of His greatest work ahead."

- Jason Dees, Senior Pastor, Christ Covenant Church

"If you desire a book that will strengthen your faith, challenge your perspective, and inspire you to leave a lasting legacy, *One More Time* is a book you'll want to read."

- Andre Kennebrew, Director of Relationships, Lifeshape

ONE
MORE
TIME
Where Chronos & Kairos Intersect
"O God, please strengthen me just one more time."
Judges 16:28b
JOHN POSEY

Published by The Preachers Trainer Inc

ISBN: 978-1-7337959-2-0

Copy Editing and Interior Layout: Michael Brewer

Cover Design: Sommer Vega

September 2026 Edition

DEDICATION

To my band of brothers in the ***CBMC Trusted Advisor Forum***—*TJ Davis, Brett Campbell, Aaron Dillon, David Barrett, JV Moore, Adonis Woods, Archie James, Tim Jaudon, Steve Carley, Paul Egan, and Scott Friderich.*

For four years, this book lived in my heart. Without your encouragement, support, and accountability, I would not have found the focus to bring it to life. Your friendship, wisdom, and belief in me gave me the perseverance to finish this important assignment. Thank you for helping me complete what God placed on my heart.

I am deeply grateful for each of you.

TABLE OF CONTENTS

Foreword

There comes a point in life when many people begin asking a difficult question: Have I already experienced my most meaningful years, or does God still have more for me to do? It is a question that surfaces in quiet moments of prayer and reflection, after career milestones have been achieved, children have grown, and dreams have either been fulfilled or adjusted. It is also the question that author John Posey addresses with wisdom, authenticity, and hope in this remarkable new book, *One More Time*.

I have known John Posey for well over a decade, meeting him for the first time in 2014, at the International CBMC Conference in Orlando. I have long admired John's commitment to helping men discover their God-given purpose and faithfully pursue it. His life and leadership have consistently reflected a deep conviction that God never wastes a season and never stops working through people who remain willing to serve Him. Throughout this book, John reminds us of a timeless truth: Our value is not determined by our age, title, accomplishments, or stage of life. Our value is rooted in God's calling and His continuing purpose for each of us. I love where John starts with Ephesians 2:10 in his new book.

What I appreciate most about *One More Time* is its refreshing message of encouragement. In a culture that often celebrates youth and achievement, while quietly sidelining experience and wisdom, John presents a different perspective. He challenges the notion that our greatest contributions are not behind us and instead points readers toward the possibility that their most significant Kingdom impact may still lie ahead. That message is both timely and deeply needed, especially to mentor and share our wisdom with millennials and Gen Z. John's challenge is for all of us to finish strong!

This book speaks directly to men who may be wondering what comes next. Some may be approaching retirement. Others may be facing unexpected transitions, setbacks, or unanswered questions. Still others may simply feel like they have drifted from the purpose and passion they once possessed. John's message is clear and hopeful: God is not finished with you. There is still purpose to discover, wisdom to share, lives to influence, and a legacy to build.

As you read these pages, I encourage you not merely to absorb information, but to embrace the invitation that author John Posey extends. Reflect on your own journey. Consider the people God has placed in your life. Think about the next generation that needs your encouragement, mentorship, and example. Ask God where He may be calling you to step forward "one more time" in faith, obedience, and service.

My prayer is that this book will inspire you to see your future through God's eyes, rather than through the limitations of your circumstances. May it rekindle hope, renew purpose, and strengthen your resolve to finish well. If you approach this book with an open heart, I believe you will discover what John so passionately communicates—that some of God's greatest work is accomplished through those who remain available to Him, regardless of age or season.

Mark Whitacre, PhD
Vice President, Culture & Care
Executive Director, t-factor
Coca-Cola Consolidated

CHAPTER 1

WHY THIS BOOK?

One of my favorite verses in the Bible is Ephesians 2:10, which in the AMPC version reads:

"For we are God's [own] handiwork [His workmanship], recreated in Christ Jesus, [born anew] that we may do those good works which God predestined [planned beforehand] for us [taking paths which He prepared ahead of time], that we should walk in them [living the good life which He prearranged and made ready for us to live]."

That verse has shaped the way I see life.

If God is truly our Father, then He goes before us. He already has plans prepared. He already sees the road ahead. Before you were born, before you took your first breath, God had purpose in mind. In fact, I sometimes say it this way: God wanted to do something in the earth, and He looked around and there wasn't the right person to do it—so He created you.

YOU ARE NOT AN ACCIDENT. YOU ARE NOT RANDOM. YOU ARE GOD'S WORKMANSHIP— HIS MASTERPIECE. HIS WORK OF ART.

The tapestry of your life, with all its victories, failures, joys, scars, heartbreaks, disappointments, and lessons, is telling a story. And your story is part of a much larger story—God's story to the human race.

That is one of the major reasons I wrote this book.

For years, I have thought deeply about what it means to finish well. Finishing well has become one of the great themes of my life. I have always been a finisher.

I finished my martial arts journey and retired as a fifth-degree black belt master. I have been married to my wife, Sandra, since December 22, 1979. I've raised three grown children.

I've stayed in the race. Longevity matters to me. But even more than natural accomplishments, I want to finish spiritually strong. Jesus, hanging on the cross, declared, "*It is finished*" (John 19:30, NIV).

The Bible calls Him the Author and Finisher of

our faith. I believe God is calling an entire generation of older men and women to finish well—not to fade away, not to become irrelevant, not to believe the lie that their best days are behind them.

We live in a culture that is quick to cancel people. Many people feel disqualified because of failure, divorce, addiction, moral compromise, lost opportunities, sickness, trauma, regret, or shame. Some people spend their later years believing God can no longer use them.

But the Bible tells a completely different story.

Hebrews 11—the great "Hall of Faith"—is filled with imperfect people. Rahab was a prostitute. Jacob was a deceiver and manipulator. David committed adultery and murder. Moses made terrible decisions and spent years in the wilderness. Yet God still used them.

Moses was eighty years old when God said, "Now your ministry begins."

God is not intimidated by your scars.

In fact, your scars may become some of your greatest qualifications.

EVERY SCAR HAS A STORY.
EVERY BATTLE MATTERS.
EVERY PAINFUL SEASON CAN BECOME WISDOM FOR ANOTHER GENERATION.

The Bible says that the righteous will still bear fruit in old age. I believe that with all my heart. Today, nearly 10,000 baby boomers retire every single day. At the same time, millions of millennials and Gen Z leaders are searching for wisdom, guidance, stability, purpose, and spiritual fathers. There is a massive transfer of wealth taking place in our culture—but not always a transfer of wisdom.

That concerns me deeply.

If we do not intentionally pass on what we have learned, then everything we have gained is only one generation away from extinction. Imagine Jesus returning to heaven and leaving nobody behind to carry the mission forward. Christianity would have disappeared within a generation. Instead, Jesus invested His life into others who would continue the work.

That is what this season of life is about.

I believe we are living at the intersection of Chronos and Kairos.

Chronos is chronological time—the years of our lives.

Kairos is God's appointed moment—the fullness of time when heaven says, "Now."

This is not the season to withdraw from purpose.

This is the season to embrace it fully.

You may feel forgotten. You may feel sidelined.

You may think your greatest opportunities have already passed. But I believe God still has paths prepared ahead of you. There are still conversations you are supposed to have; people you are supposed to mentor; wisdom you are supposed to transfer; and assignments you are supposed to fulfill.

Your life still matters.

And by the grace of God, you can still finish well.

When you study the lineage of Jesus, you quickly realize that God has always worked through broken people.

Look at Rahab. She was a prostitute, yet she became part of the lineage of Christ. Tamar's story was filled with scandal and brokenness, yet God still brought redemption through her life. Jacob was a deceiver, a manipulator, and a trickster who lied to his father and swindled his brother. His life was messy—multiple wives, family conflict, jealousy, deception, and chaos all around him. Yet God wrestled with that man until He transformed him from Jacob the trickster into Israel, a prince who had power with God.

Think about David. An adulterer. A murderer. A deeply flawed man who made catastrophic decisions. Yet God still called him *"a man after my own heart"* (Acts 13:22, NIV).

THESE WERE NOT PERFECT PEOPLE.
THESE WERE WOUNDED PEOPLE.
MESSY PEOPLE.
COMPLICATED PEOPLE.

People with scars, failures, regrets, and stories they probably wished they could erase.

And yet God used them.

That should give all of us hope.

We live in a culture today that wants to cancel people. One mistake, one failure, one bad season, and suddenly people feel disqualified forever. Many men carry hidden shame for years. Failed marriages. Bad business decisions. Addiction. Moral compromise. Financial collapse. Broken relationships. Lost opportunities. Trauma. Abuse. Regret.

Many quietly conclude, "God can't use me anymore."

But the message of Scripture is the exact opposite.

God specializes in redeeming broken stories.

The gifts and callings of God are without repentance. God does not change His mind about purpose, simply because someone walked through failure. He is able to redeem pain, restore dignity, heal wounds, and transform even the darkest seasons of our lives into wisdom that can help others.

In fact, some of the people God uses most powerfully are those who have been broken deeply and restored greatly.

YOUR SCARS ARE NOT MEANINGLESS. YOUR PAIN IS NOT WASTED. YOUR FAILURES DO NOT HAVE THE FINAL WORD.

God wants to use the hurt, the trauma, the disappointment, and even the dismal failures of your life to accelerate the growth and development of someone else. What nearly destroyed you may become the very thing God uses to heal another person.

That is one of the great purposes of finishing well.

Not pretending we were perfect.

But allowing God to redeem our story so completely, that our lives become a testimony of His grace to the next generation.

I won't get deeply into my personal story just yet—that will come later—but I do want to say that much of the work I'm doing today through the Christian Business Men's Connection was born out of a question I began asking the Lord years ago:

"How do I finish well?"

That question has stayed with me for decades.

Even when I was in my early twenties, I watched people derail their lives through moral failure, pride, compromise, greed, and hidden sin. I saw gifted leaders lose their marriages, ministries, integrity, and influence. Some started strong but finished tragically.

As a young preacher, I remember being around ministers who outwardly appeared spiritual, yet privately lived in ways that deeply troubled me. One moment in particular stayed with me. I was standing with a group of preachers on church property, while women were leaving the service. The conversation turned crude and sexual. I remember feeling disturbed inside and finally saying, "But God is holy." One of the men looked at me and said, "You're young. One day you'll be just like us."

That statement shook me deeply.

Something inside of me resisted it with everything I had.

I remember thinking, "No. I do not want to become cynical. I do not want to lose my integrity. I do not want to normalize compromise. I do not want to spend my later years pretending on the outside, while becoming hollow on the inside."

That moment became one of many defining moments in my life.

It forced me to think seriously about what kind of man I wanted to become and, even more importantly, what kind of man I wanted to be when my life was nearing its conclusion.

Because finishing well does not happen accidentally.

You do not drift into integrity.

You do not casually arrive at spiritual maturity.

You do not wake up one day and suddenly possess wisdom, character, purity, humility, and endurance. Those things are forged over time through surrender, repentance, obedience, failure, healing, and continual dependence on God.

Over the years, that burden to finish well became stronger and stronger in my heart. Eventually, it began shaping the work I now do with men, leaders, businesspeople, pastors, and young professionals. I realized that one of the greatest needs in the world today is not simply talented leaders—it is healthy leaders who can endure, remain faithful, and pass wisdom to the next generation.

That is part of why I wrote this book.

CHAPTER 1 APPLICATION
WHY THIS BOOK?

REFLECTION & ACTION

Don't just read this chapter—allow God to search your heart.

1. WHAT PART OF YOUR STORY HAVE YOU CONSIDERED A DISQUALIFICATION?

Is there a failure, disappointment, regret, or painful season that you have believed God could never use? Ask the Lord to show you how He might redeem that part of your story for His glory and someone else's benefit.

2. ARE YOU LIVING WITH THE EXPECTATION THAT GOD STILL HAS PURPOSE FOR YOUR LIFE?

According to Ephesians 2:10, God prepared good works for you before you were born. What "good work" do you believe He is calling you to in this season of your life?

3. WHO NEEDS THE WISDOM GOD HAS GIVEN YOU?

Identify one younger person—a son, daughter, grandchild, friend, coworker, or young believer—with whom you can intentionally share your story or can invest your life over the next month.

4. Take One Step This Week

Choose one practical action:

- Share part of your testimony with someone.
- Reach out to a younger leader or family member.
- Write your life lessons in a journal.
- Begin praying daily, *Lord, help me finish well.*
- Take one step toward the assignment God has placed on your heart.

Remember

God is not finished with you. As long as you have breath, you still have purpose.

CHAPTER 2

ONE MORE TIME THROUGH A BIBLICAL LENS

In his Epistle to the Romans, the Apostle Paul gives one of the greatest revelations of purpose, calling, and redemption found anywhere in Scripture:

"*And we know that all things work together for good to those who love God, to those who are the called according to His purpose*" (Romans 8:28, NKJV).

Paul goes on to explain in chapter 8, that God foreknew us, predestined us, called us, justified us, and ultimately glorified us. He then asks a series of powerful questions:

"*If God is for us, who can be against us?*" (verse 31, NKJV).

"Who shall bring a charge against God's elect?" (verse 33, NKJV).

"Who shall separate us from the love of Christ?" (verse 35, NKJV).

The answer is clear: Nothing.

Not tribulation.

Not distress.

Not persecution.

Not failure.

Not weakness.

Not even our worst moments.

Nothing can separate us from the love of God in Christ Jesus.

This passage reveals something deeply encouraging: God already knows the beginning and the end of our story. He knows our victories, our failures, our right turns and wrong turns. He sees every mistake, every delay, every season where we miss the mark—yet He still calls us according to His purpose.

That means your failure never surprises God.

Your struggle did not cancel your calling.

Your weakness does not intimidate heaven.

God is able to work even the broken pieces of our lives into His divine purpose.

One of the clearest biblical examples of this truth is the life of Samson.

Samson began with tremendous promise. Before he was born, God marked him with a divine calling. He was set apart as a Nazirite, dedicated unto the Lord with a sacred vow. The hand of God was upon his life, and supernatural strength rested on him.

But Samson's story is also a story of compromise.

He broke the very vows connected to his calling. He became careless with his consecration. His appetites and desires repeatedly pulled him off course. Eventually, his strength left him. The Philistines captured him, gouged out his eyes, bound him in chains, and forced him into slavery.

The mighty deliverer became a broken man grinding grain in darkness.

Yet Samson's story did not end in failure.

At the lowest point of his life, blind and humiliated, Samson prayed one final prayer:

"*O God, please strengthen me just one more time*" (Judges 16:28b, NLT).

That prayer changed everything.

In one final act of strength, God empowered Samson again. Though Samson's motives were still imperfect, God used that final moment to fulfill the purpose for which Samson had been born. In his death, Samson defeated more enemies than he had during his entire lifetime.

What an incredible picture of redemption.

Samson teaches us that God's purpose can survive our failures.

He teaches us that even after compromise, bondage, and loss, God can still move through a surrendered heart.

He reminds us that one moment of divine strength can change everything.

Samson made it into the great "Hall of Faith" in the Epistle to the Hebrews, in chapter 11, not because he lived a perfect life, but because God's grace and purpose were greater than his weakness.

Maybe today you feel like you have wasted years.

Maybe you feel disqualified because of past mistakes, failures, broken vows, or missed opportunities.

Maybe you once walked strongly in purpose, but

now you feel weak, tired, or spiritually blinded by disappointment.

But hear this clearly:

God is not finished with you.

The same God, who strengthened Samson one more time, can strengthen you one more time.

One more dream.

One more prayer.

One more act of obedience.

One more season of purpose.

One more opportunity to fulfill what God placed inside you before you were born.

Your story is not over.

That's why we can be so sure that every detail in our lives of love for God is worked into something good.

One of the most powerful ways to understand the Epistle to the Romans, in 8:28, is to look at it through several translations.

The ESV says:
"And we know that for those who love God all things work together for good, for those who are called according

to his purpose."

The Amplified Bible Classic Edition expands the thought even further:

"We are assured and know that [God being a partner in their labor] all things work together and are [fitting into a plan] for good to and for those who love God and are called according to [His] design and purpose."

I love that phrase: *"fitting into a plan."*

The NLT says:

"God causes everything to work together for the good of those who love God and are called according to his purpose for them."

The NIV says:

"In all things God works for the good of those who love him, who have been called according to his purpose."

And The Message Bible puts it this way:

"Every detail in our lives of love for God is worked into something good."

What an incredible promise.

These verses reveal that God is at work even in the messy places of our lives.

Many people live under crushing condemnation because of their past. That condemnation cripples our faith. It convinces us that we missed our moment, ruined our purpose, or disqualified ourselves from God's best.

THE MISTAKES.
THE WRONG TURNS.
THE FAILURES.
THE WASTED YEARS.
THE HIDDEN SHAME.
THE THINGS WE HOPE NOBODY EVER FINDS OUT ABOUT.

And the older we get, the heavier that feeling can become.

When you're young, you can make mistakes and still feel like you have plenty of runway left. Time feels abundant. Recovery seems easier. You think, "I'll fix it later."

But as you get older, you begin to realize the runway is shorter.

You become more aware of time.

You begin thinking differently about your decisions, your purpose, your relationships, and your calling. You realize you cannot afford to waste years wandering aimlessly. You want what remains of your life to count.

That realization can either produce fear and regret, or it can produce wisdom.

The good news of Romans 8 is that God is still working.

Even now.

Even after the mistakes.

Even after the delays.

Even after seasons of compromise, fear, addiction, failure, or shame.

God is still fitting the pieces together into His purpose.

That does not mean every decision was good. It does not mean every season was God's perfect will. But it does mean that God is so sovereign, so wise, and so redemptive that He can take even broken pieces and weave them into something meaningful.

That is the story of Scripture.

And that is the story of Samson.

Samson wasted tremendous potential through compromise and impulsive living. Yet even after failure, blindness, bondage, and humiliation, God still had the final word over his life.

At the end, Samson cried out:

"*[Lord], please strengthen me just one more time*" (Judges 16:28b, NLT).

And God answered him.

That same God is still answering that prayer today.

Maybe your runway feels shorter than it once did. Maybe you wish you could go back and make different decisions. Maybe you feel behind, disqualified, or weary from the weight of life.

But God is not limited by your past.

He is still able to redeem time, restore purpose, and strengthen you one more time.

CHAPTER 2 APPLICATION

One More Time Through a Biblical Lens

REFLECTION & ACTION

Don't just admire God's grace—receive it. Ask Him what He is saying to you through this chapter.

1. Where do you need God to strengthen you one more time?

Like Samson, is there an area of your life where you feel weak, discouraged, or defeated? Take a few moments to honestly ask God for fresh strength, courage, and faith for this season.

2. Have you allowed your past to define your future?

Are there failures, regrets, or missed opportunities that have convinced you that your best days are behind you? What promise from Romans 8 reminds you that God is still working in your life?

3. What "broken pieces" might God be fitting into His purpose?

Romans 8:28 reminds us that God works all things together for good. Looking back, can you identify a painful experience that God has already begun to redeem? How might He use it to encourage someone else?

4. What is your "one more time" prayer?

Complete this sentence in your journal:

Lord, strengthen me one more time to . . .

Write whatever God places on your heart, and begin praying that prayer every day this week.

5. What step of obedience have you been postponing?

God often reveals His strength as we take the next step. What is one act of obedience you know He is asking you to take? Commit to taking that step this week.

6. Who needs to hear your story of redemption?

Think of one person who feels defeated, ashamed, or ready to give up. How can you encourage them by sharing how God's grace has sustained and restored you?

Remember

Your greatest failure does not have the final word—God does. As long as you have breath, He can strengthen you one more time.

CHAPTER 3

FROM CONDEMNATION TO RIGHTEOUSNESS

One of the greatest things that plagues people is the weight of the past:

- the mistakes,
- the sins,
- the shame,
- the missed opportunities,
- the moments you knew you should have obeyed God and didn't,
- the people you should have forgiven and didn't, and
- the times you should have turned right, but turned left instead.

And over time, all of that begins to pile up on the soul.

The Bible says:

"Beloved, if our heart does not condemn us, we have confidence before God" (1 John 3:21, ESV).

And then there is another voice—the accuser of the brethren. Satan continually works against the call of God on our lives. His message is always:

- "You're too far gone."
- "God is finished with you."
- "You failed too many times."
- "You had your opportunity."
- "There's no more purpose left for you."

That voice is called condemnation.

But condemnation is different from conviction.

Conviction is the work of the Holy Spirit. Conviction says:

"This is wrong. Turn back to God."

Conviction produces repentance, humility, restoration, and life.

Condemnation is different.

Condemnation is the crushing weight of guilt and shame over sins that have already been repented of, failures already placed under the blood, and wounds from the past that the enemy keeps trying to use against you.

Conviction draws you toward God.

Condemnation drives you away from Him.

Conviction says:

"Repent and come home."

Condemnation says:

"Stay away. God does not want you anymore."

But the Bible declares:

"There is therefore now no condemnation for those who are in Christ Jesus" (Romans 8:1, ESV).

That word "now" is powerful.

NOT SOMEDAY.
NOT AFTER YOU BECOME PERFECT.
NOT AFTER YOU UNDO YOUR PAST.
NOW.
WHY?

Because of Jesus.

Because of the blood.

Because the cross was greater than your failure.

You do not have a life sentence of shame.

The blood of Jesus established a new covenant, a new standing, and a new walk. God is not relating to you on the basis of your worst moments. He is relating to you through the righteousness of Christ.

That does not mean your past did not happen.

It means your past no longer has the authority to define your future.

And maybe that is what Samson finally discovered at the end.

Blind.

Broken.

Bound.

Yet still able to pray:

"Lord . . . one more time."

And God answered him.

Which means failure did not have the final word.

Grace did.

So today:

YOU ARE NOT CONDEMNED.
YOU ARE NOT DISQUALIFIED.
YOU ARE NOT ABANDONED.

God has declared you righteous through Christ.

And if your heart has been speaking condemnation over your life, maybe it is time for your consecration to grow again.

Maybe it is time for your confidence to grow again.

Maybe it is time for your hope to grow again.

Because God still answers people who cry out:

"Lord . . . one more time."

The Story of Peter

One of the greatest lies Satan tells men in the second half of life is this:

"Because of what you've done, God is finished with you."

Many men over fifty-five carry the weight of past failures, broken relationships, bad decisions, moral failures, financial mistakes, missed opportunities, and regrets. The Bible calls Satan *"the accuser of our brethren"* (Revelation 12:10, KJV). His strategy has never changed. He continually reminds us of our past, while God continually points us toward our future.

The world operates on a system of permanent records. If a person commits a crime, serves his sentence, and pays his debt to society, the record often remains. Years later, someone can still look up the offense and identify him by what he did.

Many people mistakenly assume God works the same way.

But He doesn't.

The Bible says that when God forgives, He completely forgives. He declares:

"Their sins and their lawless deeds I will remember no more" (Hebrews 10:17, NKJV).

God doesn't merely overlook sin. He removes it. Through Christ, there is complete remission of sins. "*If we confess our sins, he is faithful and just to forgive us our sins and cleanse us from all unrighteousness*" (1 John 1:9, ESV).

The cross was not merely an emotional act of mercy. It was a legal act of justice.

Jesus became our Advocate before the Father.

"If anyone sins, we have an Advocate with the Father, Jesus Christ the righteous" (1 John 2:1, NKJV).

Satan stands as a prosecutor. Jesus stands as our defense attorney.

And no story illustrates this better than Peter.

Peter's Greatest Failure

Peter was not a weak man.

He was bold.

He was courageous.

He was the disciple who stepped out of the boat and walked on water.

He was the disciple who declared, *"You are the Christ, the Son of the living God"* (Matthew 16:16, ESV).

He was the disciple who drew a sword in the garden and swung at a soldier. Peter wasn't aiming for the man's ear—he missed his head. Peter was willing to fight.

So when Jesus told His disciples they would all forsake Him, Peter immediately objected.

"Not me."

"Even if everyone else falls away, I won't."

"Even if I must die with you, I will never deny you" (Matthew 26:35, NET Bible).

Yet Jesus looked at Peter and said,

"Before the rooster crows tomorrow morning, you will deny three times that you even know Me" (Luke 22:34, NLT).

That very night, Peter did exactly what Jesus predicted.

NOT ONCE.
NOT TWICE.
THREE TIMES.

The final denial became so intense, that Peter began calling down curses upon himself, insisting that he did not know Jesus.

Then something happened that must have pierced Peter's heart.

Jesus looked directly at him. Luke records that their eyes met.

At that moment, Peter remembered the words Jesus had spoken.

The Bible says he went out and wept bitterly.

Can you imagine the shame?

Can you imagine the disappointment?

Can you imagine the self-condemnation?

Peter had failed publicly.

He had failed completely.

He had failed at the very moment he thought he was the strongest.

I suspect Peter concluded that he was no longer qualified to be a disciple.

I suspect he believed his ministry was over.

I suspect he thought God was finished with him.

Many men today feel exactly the same way.

The Two Most Beautiful Words

After the resurrection, an angel gave a message to the women at the tomb:

"Go, tell his disciples—and Peter . . ." (Mark 16:7, NIV).

And Peter.

Why mention Peter separately?

Because Jesus knew Peter probably no longer considered himself one of the disciples.

Peter may have disqualified himself, but Jesus had not.

Those two words reveal the heart of God.

And Peter.

Not forgotten.

Not abandoned.

Not replaced.

Not disqualified.

Jesus wanted Peter to know that his failure was not the end of his story.

And neither is yours.

Restoration and Purpose

Later, Jesus met Peter on the shore while the disciples were eating fish.

Three times Jesus asked:

"Peter, do you love Me?" (John 21:15, 16, 17, GNT).

Three denials.

Three opportunities to reaffirm his love.

Jesus wasn't rubbing Peter's nose in his failure.

He was restoring him.

Then Jesus recommissioned him:

"Feed my sheep" (John 21:17, NIV).

In other words:

"I still have work for you to do."

Earlier Jesus had told Peter:

"I have prayed for you, that your faith should not fail; and when you have returned to Me, strengthen your brethren" (Luke 22:32, NKJV).

Notice Jesus did not say "if" you return.

He said "when."

Jesus already saw restoration on the other side of Peter's failure.

And the very failure that almost destroyed Peter became the foundation of his ministry to others.

One More Time

Peter's story is not about failure.

It's about restoration.

It's about grace.

It's about a God who sees beyond your worst moment.

Many men have placed themselves in what I call "spiritual hospice care." They believe their useful years are behind them. They assume God has moved on to someone younger, stronger, smarter, or more successful.

But Peter's story says otherwise.

God was not finished with Peter.

And God is not finished with you.

Your greatest contribution may still be ahead of you.

Your obedience today may impact people you'll never meet, including generations not yet born.

Peter got up one more time.

And because he did, billions of people have heard his story.

The same God who restored Peter is still restoring people today.

The question is not whether you've failed.

The question is whether you'll trust God's grace enough to get up and follow Him one more time.

CHAPTER 3 APPLICATION

FROM CONDEMNATION TO RIGHTEOUSNESS

REFLECTION & ACTION

Don't simply agree with these truths—allow them to reshape the way you see yourself.

1. WHICH VOICE HAS BEEN SHAPING YOUR LIFE?

When you think about your relationship with God, do you hear the voice of condemnation or the voice of righteousness? What lies have you believed about yourself that God's Word says are no longer true?

2. DO YOU TRULY BELIEVE YOU ARE RIGHTEOUS IN CHRIST?

It's one thing to know you've been forgiven; it's another to live as someone who has been made righteous. What would change in your daily life if you fully believed you could stand before God without guilt, shame, or condemnation?

3. ARE YOU LIVING WITH A SIN CONSCIOUSNESS OR A RIGHTEOUSNESS CONSCIOUSNESS?

When you fail, do you immediately focus on your weakness, or do you remember Christ's finished work on your behalf? How can you intentionally renew your mind to align with your new identity?

4. Is there an area where condemnation has been keeping you from God's purpose?

Identify one place where shame, regret, or feelings of inadequacy have caused you to hesitate. What step of faith can you take this week, knowing that your acceptance with God is based on Christ's righteousness—not on your performance?

5. Who needs to hear this message?

Think of someone who lives under the weight of guilt or failure. How can you encourage them this week with the truth that, in Christ, they are forgiven, accepted, and loved?

6. Declare the Truth

Read these statements aloud each day this week:

I am the righteousness of God in Christ.

I have been forgiven and fully accepted.

There is no condemnation for those who are in Christ Jesus.

I will not allow my past to define my future.

By God's grace, I will walk confidently in the purpose He has prepared for me.

Remember

You are not fighting for God's acceptance—you are living from it. Because of Jesus Christ, you can lift your head, walk in freedom, and fulfill God's purpose for your life **. . . one more time.**

CHAPTER 4

FROM CONDEMNATION TO RIGHTEOUSNESS, PART 2

The Difference Between an Old Covenant Mindset and a New Covenant Reality

One of the greatest struggles I see among men is not that they have sinned.

It is that they still think like sinners.

Many have genuinely received Christ. They have been forgiven. They have been born again. Yet they continue to relate to God as though they were living under the Old Covenant.

They carry a constant awareness of failure.

A constant awareness of shortcomings.

A constant awareness of not measuring up.

They live with what I call a **sin consciousness** rather than a **righteousness consciousness.**

The Old Covenant Produced Sin Consciousness

Under the Old Testament, the Day of Atonement was repeated every year.

Why?

Because sin had not been removed.

It had only been covered.

Every year the high priest entered the Holy of Holies with fear and trembling. He first offered a sacrifice for himself, and then for the people.

The blood covered sin, but it did not remove it.

The sacrifice had to be repeated again and again because the conscience of the worshiper was never completely cleansed.

Every sacrifice was a reminder:

"You still have a sin problem."

"You still fall short."

"You still need another sacrifice."

The entire system pointed toward humanity's inability to save itself.

The Shadow and the Substance

The Old Testament was filled with pictures of what Jesus would ultimately accomplish.

The spotless lamb.

The scapegoat.

The mercy seat.

The high priest.

The blood.

All of these pointed toward Christ.

The scapegoat carried away the sins of the people.

The lamb shed its blood for the people.

The high priest entered God's presence on behalf of the people.

Jesus fulfilled every one of those pictures, through His blood.

He became the spotless Lamb.

He became the scapegoat.

He became the High Priest.

And He offered His own blood before the throne of God.

The book of Hebrews tells us that Christ entered the heavenly Holy of Holies once and for all, obtaining eternal redemption.

NOT ANNUAL REDEMPTION.
NOT TEMPORARY REDEMPTION.
ETERNAL REDEMPTION.

The Torn Veil

When Jesus died, something remarkable happened.

The veil of the temple was torn from top to bottom.

God Himself removed the barrier.

The message was clear:

Access granted.

The way is open.

No longer would men approach God through repeated sacrifices.

No longer would they stand at a distance.

No longer would they come in fear.

Now they could come boldly to the throne of grace.

Not because of their performance.

But because of Christ's performance.

A Cleansed Conscience

Perhaps the greatest miracle of the New Covenant is not merely that our sins were forgiven.

It is that our consciences were cleansed.

Hebrews says that the blood of Christ purges our conscience from dead works.

In other words, God does not simply change our record.

He changes our standing.

He changes our identity.

He changes how we see ourselves.

Many believers know they are forgiven.

Few truly believe they are righteous.

What Is Righteousness?

Righteousness is not a reward for good behavior.

It is not spiritual achievement.

It is not earning God's approval.

Righteousness is a gift.

It is the ability to stand in the presence of God without a sense of guilt, shame, condemnation, or inferiority, as if sin never existed.

Not because you never sinned.

But because Jesus took your sin and gave you His righteousness.

"He made Him who knew no sin to be sin for us, that we might become the righteousness of God in Him" (2 Corinthians 5:21, NKJV).

Notice the verse does not say we are becoming righteous.

It says we have become righteous.

This is not something we are trying to achieve.

It is something we have received.

Two Voices

Every day we hear one of two voices.

The voice of condemnation says:

- You're not enough.
- You're a failure.
- You missed your opportunity.
- God can't use you anymore.
- You're too old.
- You've made too many mistakes.
- You'll never recover.

The voice of righteousness says:

- I am accepted in Christ.
- I am forgiven.
- I am loved.
- I am able through Christ.
- I am more than a conqueror.
- God is for me.
- Nothing can separate me from His love.

- My best days of usefulness are still ahead.

One More Time

The enemy wants you to identify with your failure.

God wants you to identify with His Son.

The enemy wants you looking backward.

God wants you looking forward.

The enemy wants you focused on what you've done.

God wants you focused on what Christ has done.

You are not the sum total of your mistakes.

You are the righteousness of God in Christ.

Therefore stand up.

Lift your head.

Receive His grace.

Walk in His righteousness.

And believe that God still wants to use you—**one more time.**

Chapter 4 Application

From Condemnation to Righteousness, Part 2

REFLECTION & ACTION

Don't simply read about your righteousness—begin living from it.

1. What Has Been Shaping Your Identity?

When you think about yourself, do you identify more with your failures or with who God says you are in Christ? What would change if you truly believed you are the righteousness of God?

2. Are You Trying to Earn What Jesus Has Already Provided?

In what ways do you still find yourself trying to earn God's love or approval through your performance? How can you begin resting in the finished work of Christ instead?

3. What Voice Are You Listening To?

Identify one recurring thought of condemnation that has followed you. Now replace it with a truth from Scripture. Commit to speaking that truth over your life every day this week.

4. Are you approaching God with confidence?

The veil has been torn, and the way into God's presence has been opened. Do you approach Him boldly as a beloved son, or do you still keep your distance because of guilt or shame? What practical step can you take to enjoy greater intimacy with God this week?

5. Where has condemnation been limiting your purpose?

Is there a ministry, relationship, opportunity, or act of obedience you've avoided because you felt unworthy? Take one courageous step this week, trusting that Christ's righteousness—not your performance—qualifies you to serve Him.

6. Help Someone Else Find Freedom

Who do you know that is living under guilt, shame, or feelings of inadequacy? Ask God to give you an opportunity this week to share the hope of the Gospel and remind them that, through Jesus Christ, they can stand before God forgiven, accepted, and free.

Remember

"One More Time" Declaration

I am not defined by my past. I am not accepted because of my performance. I am the righteousness of God in Christ Jesus. I will reject the voice of condemnation, receive God's grace, and boldly walk in the purpose He has prepared for my life . . . one more time!

CHAPTER 5

My Personal Journey to This Moment

As I begin sharing my life story, I am reminded of Esther. Mordecai told her that perhaps she had come into the kingdom *"for such a time as this"* (Esther 4:14, NIV). Most people focus on that moment; but before Esther ever arrived there, an entire story had already unfolded.

The same is true of Moses. We often think about Moses standing before Pharaoh and leading Israel out of Egypt. But before that, he spent years in the wilderness. He lost confidence in himself and even argued with God about his ability to speak. Yet earlier, the Bible describes him as mighty in word and deed.

The same pattern appears in the life of David. Before he became king, he was a shepherd tending sheep. Later, he became Saul's armor bearer. Then he became king over one tribe, and eventually king over all Israel.

Joseph had a similar story. Before the palace there was a pit, slavery, and prison. God shaped him long before He positioned him.

When I look at Scripture, I see that God often shapes people through a process, before bringing them into their purpose. My story is no different.

My story really begins with my mother.

She was thirty-three or thirty-four years old and living in the projects of Cincinnati, OH. She was on government assistance. Her husband was in jail. She already had five children. This was certainly not where she had hoped to be in life.

One day, she was sitting in church when the preacher made a statement that changed her life.

"You don't have to live in the basement of your life."

That one statement impacted her deeply. She decided she wanted something different for herself and her children, so she enrolled in nursing school.

While she was in nursing school, my biological father was released from prison. They became involved, and she became pregnant with me. She later told me that he had shown her a cut condom. In her mind, it felt as though someone was trying to stop her progress, just as she was beginning to move forward.

That's how I was born.

I've often said that I was born to stop growth, yet my purpose in life became helping people grow.

My mother was devastated by the pregnancy. She even considered abortion. Later, she told me that she threw herself down a flight of stairs hoping to miscarry, but it didn't happen.

I was born in 1959.

Now my mother had six children, and once again, she found herself alone.

She graduated from nursing school and began working as a nurse. But life remained extremely difficult. One day she became so discouraged that she put her head in the oven, intending to take her own life.

At that moment, God intervened.

She sensed Him speaking to her.

"What are you doing?"

Then another question came:

"Who's going to take care of your children?"

She answered, "What am I supposed to do?"

And the Lord replied, "Use what you've got."

That became a turning point in her life.

My mother went back to work and poured herself into providing for her family. She had an incredible work ethic. Eventually, one of the doctors she worked with noticed her diligence and dedication and asked her to care for his father through private-duty nursing.

That opportunity became the pathway that helped her get our family out of the projects.

She worked long hours—sometimes sixteen-hour days—doing whatever was necessary to provide for her children. Her sacrifice changed the trajectory of our lives. Eventually, she was able to purchase a three-story house and move us out of that environment.

It changed our lives.

Because my mother worked so much, I didn't see a great deal of her during my early years. For the first eight years of my life, I was largely raised by my brothers and sisters.

I remember moving from downtown Cincinnati, where we lived at 555 Lincoln Park Drive, to 9 Glen Este Place in Clifton. That move represented more than just a change of address. It represented my mother's determination, sacrifice, and refusal to quit.

Before all of that happened, however, my mother had a concern.

She knew that because she was working such long hours, I would spend much of my childhood in the care of my brothers and sisters. Wanting God's protection over my life, she went to church one day and asked someone to let her into the sanctuary.

She carried me to the altar and placed me there before God.

I've often joked that when your mother puts you on the altar, whatever plans you thought you had for your life have probably been rearranged. At that point, you're going to end up doing whatever God wants you to do.

My mother placed me on the altar and asked God to watch over me and protect me. For she said: "Lord, I am going to get my family out of this environment."

Then she took me home and went back to work.

From that point forward, I was largely raised by my brothers and sisters. They protected me. They looked after me. Through them, and through many circumstances throughout my life, God protected me, as well.

That's how my story began.

I began with a mother who dedicated me to God.

God watched over me through my brothers and sisters. He protected my life through countless situations I didn't fully understand at the time. Looking back now, I can see His hand on my life from the very beginning.

That's how I started.

In Martial Arts

I started with a woman who dedicated me to God. I was even called the "Gospel Choir Baby." I was one of those children who had been dedicated to the Lord, and throughout my life, the Lord took care of me through my brothers and sisters.

When I was ten years old, my mother remarried. That brought our family to fifteen children altogether, because my stepfather had nine children and my mother had six. Overnight, we became a family of seventeen.

Around that same time, my brothers introduced me to martial arts. They took me to a martial arts school, and I immediately took to it. I learned quickly and seemed to have a natural ability for it. I enjoyed the discipline, the challenge, and the structure.

I continued training until I had to leave, because there was a shootout at the location, which was called Rockdale Temple.

Later, my older sister Sheila (She was like my mom; we were very close.) encouraged me to get involved in martial arts again. She felt I needed something productive to do during the summer and wanted to keep me out of trouble. Once again, she was looking after me.

I resumed training in Pleasant Hill, and quickly rediscovered how much I loved martial arts. I became very proficient and thoroughly enjoyed it.

Eventually, my older brother Michael started his own martial arts school. I joined him there and became an assistant instructor. I was only thirteen years old. Michael was like a father to me. He taught me not only martial arts. (There is a very special relationship between a Sensei and student.) But, he served as a towering picture of what it means to be a man.

Looking back, that experience became one of the first leadership laboratories of my life.

As an assistant instructor, I was responsible for teaching incoming students.

People of all ages would come through the school, and my job was to teach them the fundamentals of martial arts. Beyond teaching techniques, I helped students discover what they did well, and encouraged them to develop their strengths.

Interestingly, that's still much of what I do today. Whether I'm working with leaders, young professionals, ministry teams, or individuals, I'm still helping people learn the basics of life, spiritual growth, leadership, and successful living. Then I help them discover what they do well and encourage them to grow in those areas.

That's what I was doing then, and it's what I continue to do now.

Through martial arts, I learned responsibility, authority, leadership, respect, discipline, and focus. Those lessons would stay with me for the rest of my life.

At sixteen years old, I earned my first-degree black belt. At that time, I was the youngest black belt in my association and was respected among those ranks.

Martial arts continued to play an important role in my life for many years. Eventually, I achieved the rank of fifth-degree black belt master.

Chapter 5 Application
My Personal Journey to This Moment

Reflect
Looking back on one of the most difficult seasons of your life, what strengths, character traits, or lessons was God developing in you that you couldn't see at the time? How might those experiences be preparing you for your next assignment?

Share
Is there someone in your life—a child, grandchild, younger leader, or friend—who needs to hear one of your "wilderness" stories? What experience from your journey could encourage them to keep trusting God through their own process?

Encourage
Think of a time when someone saw potential in you before you saw it in yourself. How did their belief change the course of your life? Who has God placed in your life today who needs you to see, affirm, and call out the potential in them?

CHAPTER 6

My Calling to Gospel Ministry

I received Jesus Christ into my life in the summer of 1974; I was fifteen.

Shortly after that, I began reading the Bible. I believed what I read, and as I read the Scriptures, I discovered that Christians were supposed to share their faith with others. So that's exactly what I started doing.

I began sharing the Gospel, and people responded.

Students started coming to Christ, and before long, I found myself gathering together groups of young people and teaching them what I was learning. In many ways, I started discipling others almost immediately after becoming a believer myself.

I remember asking God to give me the ability to teach.

And He did.

As I continued growing in my walk with the Lord, opportunities to share my faith continued to increase. Eventually I found myself preaching at my school.

That didn't go unnoticed.

At one point, the principal, Mr. Henderson, called me into his office and informed me that I was disrupting the school. He told me there was a separation of church and state and that I needed to cease and desist. That night, I went home and began reading the Book of Acts. I came across the story of Peter and John being threatened and commanded not to preach the Gospel.

> Peter's response captured my attention. *"We ought to obey God rather than men"* (Acts 5:29, NKJV).

As I read those words, I made a decision.

I wasn't going to stop preaching!

The next day, I returned to school. I gathered with my preaching partner, Mark Toney, and a few of the new disciples. I told them, "I'm going to preach the Gospel again."

"No, Johnny, don't do it!" they pleaded. "You'll get in trouble."

I looked at them and said, "No. I must obey God."

I walked out to the smoking patio, where students gathered between classes. As they began to assemble, I stood up and started preaching Jesus. I wasn't trying to be original. I simply took the sermons recorded in the Book of Acts and proclaimed the same Gospel the apostles preached.

Tears streamed down my face as I cried out:

"Jesus is both Lord and Christ! Repent! Turn away from this wicked and perverse generation, and your sins will be forgiven!"

I wasn't concerned about what people thought of me. In that moment, I was overwhelmed with one desire—to make Jesus known. The fire of God burned so deeply in my heart, that I couldn't remain silent.

The response was immediate.

Seventeen students responded to the Gospel that day.

Before long, I found myself back in the principal's office.

He looked at me and said, "I told you not to preach."

I replied, "I'm not going to stop."

Then he said, "If you continue doing this, I'm going to expel you from school."

My response was simple.

"I would rather obey God than man."

Looking back, I can see how sincere and determined I was in my walk with the Lord. I wasn't trying to rebel against authority. I simply believed that God had called me, and I wanted to be obedient.

Mr. Henderson sat with his head in his hands. He simply didn't know what to do with me.

Finally, he asked, "Johnny, what church do you belong to, and who is your pastor?"

I replied, "St. John AME Zion Church. Rev. Percy Smith is my pastor."

Mr. Henderson looked up the church's phone number and called him.

"Rev. Smith," he said, "I have one of your members here, Johnny Posey. You need to talk to him about preaching at my school. He's disrupting things around here."

When Rev. Smith came to the phone, he didn't tell me to stop preaching. Instead, he said, "Johnny, render to Caesar the things that are Caesar's, and to God the things that are God's. Don't stop preaching—but you need to change your strategy."

"Yes, Pastor," I replied.

I wasn't expelled that day, but I did go underground.

I kept sharing the Gospel, but only in places where I wasn't violating school rules. I discovered I could preach on the sidewalk outside the school, while students waited for their buses. I preached on the school bus rides home. Anywhere I had an opportunity, I talked about Jesus.

One afternoon, while riding the bus home, the Word of God was burning so strongly in my heart that I could no longer keep quiet. I stood up and boldly proclaimed, "Hear me! Jesus is Lord, and He is coming again! Repent of your sins and believe the Gospel!"

More than forty years passed.

Then, about eight years ago, I received a message on Facebook. The first thing that caught my attention was how the man addressed me.

"Johnny."

I hadn't been called that name in decades.

I replied, "Do I know you?"

He answered, "No, but I was on the school bus, where you preached the Gospel over forty years ago."

Then he shared something that brought tears to my eyes.

"Today I'm a bishop serving in ministry. Two other friends, who were on that same bus, are also in

Gospel ministry. We wanted you to know that your boldness that day impacted our lives."

I sat there overwhelmed.

What I thought was a simple act of obedience had become part of God's eternal story.

That experience taught me something I have never forgotten:

> You rarely know what God is doing through your obedience. Sometimes the seeds you plant disappear into the ground for decades, before you ever see the harvest. But God never forgets an act of faith. Every conversation, every witness, every act of courage matters, even when you never see the immediate results.

Looking back, I can see that the Lord was already stirring my heart toward a lifetime of Gospel ministry. What began as a teenage passion to tell others about Jesus was His way of preparing me for the calling that would shape the rest of my life.

I remember one particular day very clearly. It was the summer of 1975, I was walking to my summer job, and tears were streaming down my face.

As I walked, I prayed.

"Lord, I will do whatever You call me to do. Whatever You want me to do. I want to serve You."

In my heart, I sensed His response.

"I want you to preach My Gospel."

That answer created a tremendous sense of responsibility within me.

To be honest, I wasn't sure I wanted that responsibility. The thought of being responsible for people felt overwhelming. For a while, I tried to ignore it. As much as I could, I pushed it aside. But eventually, I knew I had to answer.

I said, "Yes."

I went to my pastor and told him what I believed God was saying. He instructed me to spend time praying about it, which I did.

After about two months, I returned and told him, "I believe I'm called to preach the Gospel of Jesus Christ."

At the time, I was only fifteen years old and getting ready to turn sixteen.

My pastor scheduled me to preach what was called a trial sermon. The date was December 31, 1975, and this would be my first opportunity to preach publicly.

As always, my family stood behind me.

My older brother Michael helped me prepare. In

fact, all of my brothers and sisters were incredibly supportive throughout my life. Looking back, that support helps explain why I care so much about developing young people today.

The confidence I had, the belief I had in myself, and the support system surrounding me created a sense that I could accomplish anything God called me to do.

Michael worked with me on my message.

The text I chose was 1 Corinthians 10:31:

"Whether therefore ye eat, or drink, or whatsoever ye do, do all to the glory of God" (KJV).

I built my sermon around the theme of living for the glory of God.

My brother helped me learn how to tell stories, illustrate points, and communicate effectively. Before the service, he gave me a small cross that belonged to him.

He placed it around my neck and said, "I want you to wear this when you preach."

I wore that little cross as I stood to deliver my first sermon.

What happened next is difficult to describe.

I remember feeling as though I had stepped outside myself. I had never experienced the power of God quite like that before.

As I preached, I felt the power of God come upon me. It was almost as though I was watching myself from the outside. I found myself preaching about Jesus, about how much I loved Him, and about how every aspect of our lives should be lived for the glory of God.

Whether we eat or drink, whatever we do, we are to do it for God's glory.

We are to glorify Him with our lives. Interestingly, that's how I began, and that's how I remain today.

My desire is still the same.

I want to live for God's glory.

I want my life to glorify Him in every way.

Today we often use the term "image bearer." We are called to bear the image of God in the earth.

That is still my desire.

I want to be an image bearer.

I want to represent Him well in every area of my life.

Sandra, Ministry, and Learning to Measure Success Differently

As my life began to unfold, I found myself walking two distinct paths. One path was my professional career in engineering, computers, technology, and business. The other was my spiritual journey—the path that would ultimately define my life's work and purpose.

It was during my college years, around the age of nineteen, that I began developing a relationship with Sandra, the woman who would eventually become my wife. Looking back now, I realize how significant that relationship was to everything that followed. Finding the right person to share your life with is one of the most important decisions you will ever make. For many people, life feels as though it is on hold while they search for a spouse, a covenant partner, someone with whom they can build a future. I was fortunate to find that person early in life.

Sandra and I began a journey together that has now lasted more than four decades. Like every couple, we had to "learn" each other. We learned how to walk together, how to share our lives, how to dream together, and how to work through the issues that inevitably arise in any meaningful relationship.

What I eventually discovered was that my relationship with Sandra became the DNA for every other relationship in my life. The skills I learned in marriage—communication, forgiveness, unity, honesty, authenticity, and integrity—became the foundation for leadership, ministry, business, and every other area of life.

During those early years, I had concerns about ministry. I had seen enough to know that not everyone finished well. I had witnessed moral failures, false doctrines, broken relationships, and people losing their way. I wondered whether I would be able to avoid those same pitfalls.

It was during that season that I sensed the Lord speaking something profound to me.

The Lord showed me that I could go out and accomplish some things on my own. I could do reasonably well by myself. But if I learned how to come into agreement with Sandra, if we learned how to truly walk together in unity, we could accomplish exponentially more.

I call it supernatural math.

One plus one does not equal two.

One plus one equals ten.

The Lord showed me that together we could accomplish ten times more than either of us could accomplish alone.

At the same time, He redirected my ambitions.

Rather than pursuing success, fame, or fortune, I was to pursue mastery of the love walk. My first ministry assignment was not a church. It was not a platform. It was not a congregation.

It was learning how to love Sandra.

Those principles became foundational to my life. I learned that love involves forgiveness. It requires working through conflict. It demands honesty, authenticity, humility, and integrity. It means valuing another person and seeking unity even when it is difficult.

Many of the lessons I learned during those early years with Sandra became lessons that carried me through every subsequent season of life.

As my spiritual journey continued, I attended Bible school and immersed myself in studying Scripture and theology. I began serving in a local church, eventually becoming deeply involved in church life.

Over time, I was hired as an assistant pastor.

That opportunity became one of the greatest learning experiences of my life.

I learned about church administration, organization, goal setting, and time management. More importantly, I learned about people.

I learned how to work with staff members, church leaders, volunteers, and congregants. I learned patience. I learned long-suffering. I learned how to navigate conflict and disappointment. I learned how to serve faithfully even when things were not going my way.

During that season, I developed relationships that remain meaningful to me today.

One of those relationships was with Tim Mann, who eventually became the pastor of the church. Tim was an important mentor in my life. He taught me how to solve problems, how to work through difficult situations, and how to listen to the voice of the Lord.

He also helped me grow in my understanding of the Holy Spirit. Tim was deeply charismatic and Spirit-filled, and through his influence, I developed a richer relationship with the Holy Spirit and a greater sensitivity to God's leading.

As the years passed, Sandra and I sensed a call from God to move to Atlanta, GA, and plant a church.

Our journey had taken us from Cincinnati to Tulsa. Now it would take us to Atlanta.

There we founded Zoe Christian Fellowship.

The word "Zoe" speaks of the life of God, and that became the foundational principle of the church. We poured our hearts into building that ministry. Over time, I pastored the church for more than thirty years.

Those years were filled with both tremendous blessings and tremendous challenges.

Like every pastor, I dealt with people problems, family issues, staff conflicts, financial pressures, church discipline situations, disappointments, betrayals, and painful experiences.

There were seasons of growth and seasons of decline.

I experienced the highs and lows of ministry.

Throughout those years, I carried a dream.

I dreamed of building a large church—perhaps even a megachurch.

It was a dream that never materialized.

Several times we would grow, only to experience setbacks. People would leave. New challenges would arise. We would rebuild, only to face another obstacle. The cycle repeated itself over and over again.

I attended church growth conferences. I studied leadership. I invested in training. I implemented strategies. I worked tirelessly.

Yet, the vision I had imagined never became reality.

For a long time, I viewed that as failure.

When I eventually stepped away from pastoral ministry, I felt as though I had not achieved what I had hoped to achieve.

As I grew older and became increasingly aware that I had fewer years ahead than behind, those feelings intensified.

To be completely honest, there were times I felt embarrassed.

I was embarrassed by the size of the church. I was embarrassed by the demographics. Even though we had helped many people and borne genuine fruit, I felt as though I had fallen short of what I believed I was supposed to accomplish.

I felt like I had hit a ceiling that I could never break through.

Sandra carried those burdens alongside me.

Ministry had been difficult for her, as well. The pressures, challenges, and disappointments affected both of us.

In 2012, I eventually transitioned out of pastoral ministry, and she was genuinely relieved. For her, it represented the end of a very difficult season.

That same year, after completing my John

Maxwell Certification as a trainer, coach, speaker, and DISC consultant, I launched a new ministry focused on equipping pastors, leaders, and churches. Sandra and I began traveling full-time across the country, investing in leaders and strengthening local churches. That season became one of the most fulfilling and fruitful chapters of our lives. Many of the lessons and experiences from those years deserve a chapter of their own.

Then, in 2018, something happened that I never saw coming.

As I approached sixty years old, I found myself asking a profound question:

"What do I do with the rest of my life?"

I sought the Lord deeply during that season.

What I sensed Him saying changed everything.

The Lord showed me that there were lessons I had not learned earlier in life. There were things I could not go back and fix.

There were dreams that would never be fulfilled in the way I originally imagined. I had to accept that.

I had to stop reliving what I did or did not accomplish.

I had to turn the page.

The Lord was inviting me into a new season.

Once I embraced that truth, I was able to release the past and focus on what God was saying in the present.

That new season led me to adjust my focus in my efforts in re-launching Christian Business Men's Connection (CBMC) Atlanta in 2015, and the work I am doing today.

What I thought was an ending turned out to be a beginning.

All of the disappointments, setbacks, betrayals, frustrations, failures, and painful experiences had actually been preparation. God had been shaping me through every one of those seasons.

Today, I can honestly say that this has become the most fruitful season of my life.

I am accomplishing more than I ever imagined. I am having a greater impact than I ever had before. I am investing in the next generation of leaders and helping them avoid some of the mistakes and struggles that I encountered.

What once looked like failure was actually preparation.

**THE TRIALS PREPARED ME.
THE DISAPPOINTMENTS
PREPARED ME.
THE SETBACKS PREPARED ME.
THE PAIN PREPARED ME.**

And now, because I walked through those experiences, I have something valuable to offer others.

I can help serve the next generation of leaders.

And perhaps that was God's plan all along.

Chapter 6 Application
My Calling to Gospel Ministry

Reflect
When did you first sense God calling you to something greater than yourself? How did you respond then, and how has your understanding of that calling changed over the years?

Imagine
Imagine looking back on your life twenty years from now. What would you want people to say about your faithfulness, rather than about your achievements? How would that change the way you live today?

Evaluate
How do you currently measure success? Is it by numbers, accomplishments, income, or recognition or by faithfulness and obedience? In what areas might God be inviting you to redefine success?

Examine
I discovered that my first ministry assignment was learning to love my wife well. What relationships is God using to shape your character today? What qualities—humility, forgiveness, honesty, patience, or unity—does He want to strengthen in you?

RELEASE

Is there a disappointment, unmet dream, or perceived failure that you continue to carry? What would it look like to release it to God and trust that He has been using that season to prepare you for what comes next?

INVEST

Who is the next-generation leader God has placed in your life? What is one practical way you can encourage, mentor, or invest in that person this week?

LEARN

Think about one of the most painful seasons you've experienced. Looking back, what did God teach you through it, that you could now use to help someone else?

PRAY

Ask God to reveal whether there is any area where you have confused success with faithfulness. Pray that He would give you the courage to embrace His definition of success and the wisdom to fulfill your calling with humility.

SHARE

Who do you know that feels discouraged because life hasn't turned out the way they expected? How could you encourage them this week by sharing how God can transform setbacks into preparation for a greater purpose?

ACT

Write down one step of obedience God has been prompting you to take. Stop waiting for the perfect circumstances and commit to taking that step this week, trusting that faithfulness today prepares you for tomorrow.

What if the greatest work God wants to do through your life isn't what you build—but who you become and who you prepare to carry the mission after you? How would that change the way you live, lead, love, and invest in others—beginning today?

Throughout this chapter, we've seen that God's calling is often revealed one step at a time. It is shaped through obedience, refined through relationships, strengthened by trials, and fulfilled through faithfulness. What once appeared to be failure often becomes preparation for a greater purpose.

CHAPTER 7

The Intersection of Chronos and Kairos

"But when the fullness of time had come . . ." (Galatians 4:4, ESV).

"Of Issachar, men who had understanding of the times, to know what Israel ought to do, 200 chiefs, and all their kinsmen under their command" (1 Chronicles 12:32, ESV).

"Look carefully then how you walk, not as unwise but as wise, making the best use of the time, because the days are evil" (Ephesians 5:15-16, ESV).

One of the most important lessons I've learned is that not all moments are created equal.

The Bible describes two different kinds of time.

Chronos is the passage of time. It is measured in minutes, hours, days, and years. It is the calendar on the wall, the watch on your wrist, birthdays, anniversaries, deadlines, and retirement dates.

Chronos simply keeps moving. It never stops for anyone.

But Scripture also speaks of Kairos.

Kairos is not measured by clocks.

Kairos is measured by God's activity.

Kairos is that divine moment when heaven touches earth, when God's Kingdom advances in a unique way, and ordinary men and women are invited to participate in extraordinary purposes. It is the moment when God's will is done on earth as it is in heaven. At those intersections, the impossible becomes possible because God Himself is at work.

Chronos asks,
"What time is it?"

Kairos asks,
"What is God doing?"

One measures time.

The other measures opportunity.

One records history.

The other changes history.

When heaven kisses earth, and you get caught in the smack!

I like to think of Kairos this way.

There are moments when heaven comes down and kisses the earth. For a brief season, eternity intersects with time.

God's Kingdom presses into the natural world, and ordinary people suddenly find themselves standing inside extraordinary moments. Those who recognize these moments often experience an unusual grace. Doors that remained closed for years suddenly open. Relationships form that could never have been orchestrated. Influence increases. Resources appear. There is an acceleration that cannot be explained by talent, experience, or human effort alone.

Those moments cannot be manufactured.

They cannot be scheduled.

They cannot be explained.

You simply discover that God is already moving, and you have been invited to participate.

Jesus taught us to pray,

"Your kingdom come. Your will be done on earth as it is in heaven" (Matthew 6:10, NKJV).

Whenever that prayer becomes reality, Kairos has arrived.

Throughout Scripture

The Bible is filled with these intersections.

Moses experienced one at the burning bush.

For forty years, he had lived in Chronos.

One ordinary day became extraordinary because God appeared.

David experienced one when he faced Goliath.

To everyone else, it looked like a shepherd fighting a giant.

In reality, heaven had entered the battlefield.

That single stone carried more than David's strength.

It carried God's purpose.

Elijah experienced one.

The Spirit of God came upon him, and he outran Ahab's chariot.

Natural ability was overtaken by supernatural grace.

Pentecost was another.

One ordinary feast day became the birth of the Church. History changed forever.

None of these people created the moment.

They recognized it.

Then they stepped into it.

The Intersection

The most powerful place to live is where Chronos and Kairos intersect.

Chronos provides the calendar.

Kairos provides the calling.

Chronos tells us where we are.

Kairos tells us why we are there.

Most people spend their lives simply watching the clock.

Few recognize when God has entered the room.

But when they do, everything changes.

Suddenly there is unusual favor.

Unexpected influence.

New opportunities.

Open doors.

Acceleration.

It almost feels effortless.

Not because life has become easy.

But because God is supplying grace for the assignment.

You are no longer rowing upstream.

You are moving with the current of God's purpose.

Our Generation's Kairos

As I said before, I believe we are living in one of those moments right now.

Every single day, approximately 10,000 baby boomers leave the workforce.

An unprecedented transfer of leadership, wealth, experience, relationships, and influence is taking place.

At the same time, millions of millennials and

members of Generation Z are stepping into positions of responsibility. This is more than a demographic statistic.

It is a Kingdom opportunity.

Never before have we witnessed such a massive transfer of influence between generations.

The question is not whether the transfer will happen.

It will.

The question is whether wisdom will be transferred along with it.

That is the Kairos before us.

This is why mentoring matters.

This is why discipleship matters.

This is why succession matters.

God is inviting an entire generation to participate in something much larger than themselves.

Grace for the Moment

One of the remarkable things about Kairos moments is that they carry their own grace.

There are things that you struggled to accomplish for years.

Then suddenly, during God's appointed time, doors begin opening.

PEOPLE APPEAR.
RESOURCES COME.
FAVOR INCREASES.
MOMENTUM BUILDS.

The assignment begins moving almost beyond your own ability.

It isn't because you suddenly became smarter.

Or more talented.

Or more experienced.

It is because God's grace accompanies God's timing.

The work becomes fruitful because heaven has entered the moment.

You begin realizing that God was already at work before you ever arrived.

You are simply joining Him.

You Can't Explain It

People often ask how these things happen.

The truth is, many times you can't explain them.

You can only testify to them.

Looking back, you realize that God had already arranged the people, the opportunities, the timing, and the circumstances.

You simply obeyed.

You stepped into the moment.

History was already moving.

God simply invited you to participate.

That is the beauty of Kairos.

It is not about making something happen.

It is about recognizing what God is already doing—and having the faith and courage to join Him.

I believe we are living in one of those moments today.

I believe this is a Kingdom moment.

The greatest transfer taking place is not wealth.

It is not leadership positions. It is not businesses or organizations.

It is the transfer of wisdom.

It is the transfer from spiritual fathers and mothers.

It is the transfer of faith, character, experience, and Kingdom vision to the next generation.

That is why I have dedicated my life to building leaders, developing young professionals, and encouraging, equipping, and empowering spiritual fathers. This is not simply another ministry strategy. It is my response to what I believe is one of the greatest Kairos moments of our lifetime.

Again, the question is not whether this transition will happen. It will. The question is whether God's people will recognize it.

Will we spend the final chapters of our lives protecting what we have built?

Or will we invest ourselves in those who will build what comes next?

For men and women in the second half of life, this may be the greatest invitation God has ever extended.

The first half of life was about building success.

The second half is about building successors.

The greatest privilege of our lives is not convincing God to bless our plans. It is recognizing where He is already moving and having the courage to join Him.

This is our Kairos.

May we not miss it.

Why does "ONE MORE TIME" matter?

History is standing at the intersection of Chronos and Kairos. Chronos tells us that one generation is leaving the stage, while another is stepping into the spotlight. Kairos tells us that God is doing far more than orchestrating a generational transition—He is creating a divine appointment.

Never before has there been such a massive transfer of wealth, influence, leadership, and opportunity. Yet the greatest transfer God desires is not financial; it is spiritual. He is calling this generation of spiritual fathers and mothers to pass on the wisdom, faith, character, and Kingdom perspective that can shape generations yet unborn.

This is more than a demographic shift.

It is a Kingdom succession strategy.

The window of opportunity will not remain open forever. Kairos moments always require a response.

They invite us to step beyond comfort, beyond retirement, and beyond self-preservation into God's greater purpose.

The question is not whether history is changing.

It is.

The question is whether we will recognize what God is doing before the opportunity passes.

I believe this is our moment.

This is our Kairos.

May we have the wisdom to recognize it, the faith to embrace it, and the courage to step into it.

God is calling this generation of spiritual fathers and mothers to transfer their wisdom before this unprecedented window of opportunity closes.

My Kairos Moment

In 2018, my age was 59, and I was seeking the Lord about how I would finish my life. I wasn't asking how to become more successful. I was asking, "Lord, how do You want me to finish?"

During that time of prayer, I sensed the Lord speak clearly to my heart:

"I want you to pour yourself out into the next generation. Anything that is not directly impacting men between the ages of twenty-two and thirty-nine, I want you to begin phasing them out of your life. This is what I have called you to do."

Looking back, I realize what He was really saying was, "John, I have a Kairos moment for you. Step into it."

That single direction changed the trajectory of my life.

I began restructuring everything around that assignment. I phased out many activities that had once been important to me. I closed businesses. I resigned from nonprofit leadership positions. I reprioritized my relationships. I withdrew my leadership roles in multiple ministerial organizations. Some of these roles were sources of income. Even within CBMC, I refocused our entire strategy around reaching and developing young professionals.

I wasn't asking God to bless what I was already doing. I was adjusting my life to what God was already doing.

That is the nature of a Kairos moment.

We often pray for God to join our plans, but Kairos works the opposite way. God invites us to leave our plans behind and join His. The question is never, "God, will You bless what I'm building?" The question is, “Lord, where are You moving, and how can I join You?”

When we recognize that moment and have the courage to step into it, something extraordinary happens. We become caught up in the current of God’s activity. We begin experiencing His grace in ways that cannot be manufactured through human effort.

Today, I can honestly say I am living through one of the most fruitful seasons of my life. The favor I have experienced, the relationships God has opened, the opportunities He has provided, and the resources He has entrusted to me are greater than at any previous time in my life.

It isn’t because I became smarter. It isn’t because I worked harder.

It is because I stepped into my Kairos moment.

I simply caught the current of what God was already doing.

Chapter 7 Application

The Intersection of Chronos and Kairos

Reflect

When have you experienced a moment that felt like a Kairos moment—a time when you sensed God was doing something beyond ordinary circumstances? How did you respond, and what was the outcome?

Discern

Are you primarily living by **Chronos**—watching the clock, managing your schedule, and checking off responsibilities—or are you intentionally looking for where God is already at work? What might He be inviting you to join in this season?

Release

Think about a time when you had to let go of your own plans and trust God's timing instead. What did you learn through that experience, and how has it strengthened your faith?

Invest

I describe this generation as living in a unique Kairos moment—a time when wisdom, faith, and experience can be transferred to the next generation. Who has God placed in your life to encourage, mentor, or disciple? What is one intentional step you can take this week to invest in them?

Reorder

If God asked you today to realign your life around His current assignment—as He did with me—what activities, commitments, or priorities might you need to reduce, release, or eliminate in order to fully embrace His purpose?

Pray

Ask God to give you "the understanding of the times," like the sons of Issachar. Pray for spiritual discernment to recognize where He is moving, the courage to join Him, and the wisdom to invest your life in what will have eternal significance.

Act

Write down one opportunity you believe God may be placing before you right now. Instead of asking Him to bless your plans, ask, **"Lord, where are You already working, and how can I join You?"** Then take one practical step of obedience this week.

Final Reflection

What if your greatest opportunity isn't something you need to create—but something God is already doing, that He is inviting you to join? How would your priorities—your prayers, and your daily decisions change if you truly believed that this season is your **Kairos?**

CHAPTER 8

WHY SIT HERE UNTIL WE DIE?

Let me begin with a question:

How do people settle for spiritual and emotional hospice care?

Hospice care has its place, when a person is approaching the end of physical life. Its purpose is to provide comfort when there is no expectation of recovery. But what happens when people who are still alive, still breathing, and still called by God, begin to live as if nothing can change? What happens when they stop believing, stop trying, stop growing, and simply wait for the end?

I believe many people have unknowingly entered spiritual and emotional hospice care.

The title of this chapter comes from a remarkable story found in the book of 2 Kings 1-20. Please take a moment to read this entire chapter.

The people of God were experiencing one of the darkest periods in their history. The Syrian army had surrounded the city of Samaria. The siege had lasted so long, that a devastating famine swept through the city. Food became so scarce, that donkey heads were sold for exorbitant prices, and the situation deteriorated to the point where people resorted to cannibalism. Conditions could hardly have been worse.

Outside the city gate sat four lepers.

Because of their disease, they were considered ceremonially unclean and lived separated from the community. They found themselves trapped between two impossible realities.

If they entered the city, they would find a famine-stricken population with no food to offer them. Entering the city also violated the law concerning leprosy. To go back meant certain death.

If they remained where they were, they would eventually starve to death.

Then they considered a third option: the camp of the Syrians.

The enemy camp represented uncertainty. The Syrians might kill them. They might show them mercy.

They simply did not know.

And then they asked one of the most powerful questions in all of Scripture:

"Why sit we here until we die?" (2 Kings 7:3, KJV).

That question marked a turning point.

The four lepers recognized something important: They had nothing to lose by moving forward.

Staying where they were guaranteed death.

Going back guaranteed death.

But moving toward the unknown at least created the possibility of life.

So they got up.

They took a step.

And as they moved toward the Syrian camp, God moved on their behalf. The Lord caused the Syrian army to hear the sound of horses and chariots and a great army. Fear seized them, and they fled, leaving behind their tents, food, clothing, silver, and gold.

The very place the lepers feared, became the place of God's provision.

I wonder how many of us are sitting at the gate of

our own lives asking the wrong questions.

We say:

- “What if I fail?”
- “What if I get sick?”
- “What if my wife leaves?”
- “What if people reject me?”
- “What if things don’t work out?”
- “What if I don’t have what it takes?”

But perhaps the better question is this:

Why are we sitting here until we die?

Why are we remaining in bitterness until our relationships die?

Why are we remaining in fear until our dreams die?

Why are we remaining in complacency until our purpose dies?

Why are we settling for spiritual and emotional hospice care, when God is calling us to get up and move forward?

You may not know what lies ahead.

You may feel trapped between an uncomfortable past and an uncertain future.

But one thing is certain: **Remaining where you are will not produce the life God intends for you.**

The lepers did not have a perfect plan.

They did not have certainty.

They simply decided that staying put was no longer an option.

Sometimes the greatest act of faith is simply getting up.

Because when you take a step of obedience, God can do what only God can do.

The question remains:

Why are you sitting here until you die?

Chapter 8 Application

Why Sit Here Until We Die?

Reflect

What area of your life has quietly slipped into "spiritual hospice care"? Where have you stopped believing, stopped growing, or stopped expecting God to work?

Examine

What fear has kept you sitting at the gate? Is it fear of failure, rejection, criticism, disappointment, or simply the fear of the unknown?

Count the Cost

The four lepers realized that staying where they were, guaranteed death. What is—remaining where you are—costing you spiritually, emotionally, relationally, or in fulfilling God's purpose for your life?

Step Forward

Is there a "Syrian camp" in your life—an opportunity, conversation, ministry, career move, or act of obedience you've been avoiding? What is one practical step you can take this week, even without knowing the outcome?

Remember

Think of a time when you stepped out in faith, despite uncertainty. How did God meet you? What does that experience teach you about trusting Him today?

Believe

What "impossible" situation have you concluded will never change? How might God be inviting you to trust His power instead of your circumstances?

Pray

Ask God for the courage to leave the place of comfort and complacency. Pray that He will replace fear with faith, give you discernment for your next step, and help you trust Him with the outcome.

Act

Write down one area where you have been waiting instead of obeying. Then identify one specific action you will take this week. Don't wait until you have all the answers. Simply take the next step and trust God to meet you there.

Final Reflection

The greatest risk is often not taking the wrong step—it is taking no step at all. If you knew God was already preparing the way ahead of you, what would you stop postponing? What would you do today instead of sitting at the gate?

CHAPTER 9

HOPE: THE FUEL FOR ONE MORE TIME

"May the God of hope fill you with all joy and peace as you trust in him, so that you may overflow with hope by the power of the Holy Spirit" (Romans 15:13, NIV).

There is something every man needs to keep moving forward.

IT ISN'T MONEY.
IT ISN'T HEALTH.
IT ISN'T TALENT.
IT ISN'T INFLUENCE.
IT IS HOPE.

Hope is the fuel of the human spirit.

When hope dies, people stop dreaming. They stop believing. They stop trying. They may still go through the motions of life, but something inside them has quietly shut down.

I've met many men who weren't physically dead, but emotionally and spiritually they had stopped living years earlier. They simply existed.

They had lost hope.

That is one of the reasons I wanted to write this book.

A Nation Looking for Hope

Research tells an interesting story.

Most Americans remain hopeful about their own future and the future of their families. Yet far fewer are hopeful about the future of our nation. We live in a culture marked by uncertainty, division, economic pressure, and rapid change.

For men, the struggle can be even deeper.

Men die by suicide at significantly higher rates than women, with the highest rates occurring among middle-aged and older men. Hopelessness, isolation, financial distress, health challenges, and relationship loss all contribute to this heartbreaking reality.

Many men have spent decades building careers, raising families, and providing for others. Then one unexpected event—a layoff, a divorce, a diagnosis, retirement, or simply growing older—causes them to question who they are.

Who am I if I no longer have my title?

Who am I if my children are grown?

Who am I if I can no longer do what I used to do?

Research consistently shows that men over forty wrestle with fears of losing their identity, financial security, relevance, health, and purpose.

Those fears are real.

But they don't have to define us.

Hope Is Like Fuel

Imagine driving down the interstate.

Suddenly your car begins to sputter.

The engine loses power.

Eventually the vehicle rolls to a stop.

What happened?

The engine didn't disappear.

The transmission still works.

The tires still have tread.

Nothing may actually be broken.

The problem is much simpler.

The car has run out of fuel.

Many men are living exactly that way.

They still possess wisdom.

They still have gifts.

They still have decades of experience.

God has not taken away their calling.

They've simply run out of hope.

Hope is what gets you out of bed when yesterday disappointed you.

Hope gives you the courage to apply for another job.

Hope gives you the strength to apologize after failure.

Hope gives you the determination to begin again.

Hope whispers three simple words:

One more time.

When Life Forces You to Pivot

Many men never imagined their lives would look the way they do today.

Some expected to retire comfortably.

Others expected to spend thirty years with one company.

Some dreamed of traveling.

Others imagined holding grandchildren, while watching the business they built continue to flourish.

Instead, life changed.

Jobs disappeared.

Companies downsized.

Health changed.

Relationships broke.

Dreams were delayed.

Workers over fifty often face significant challenges

finding meaningful employment after losing a job. Many even report feeling pushed out of the workforce because of age.

Maybe that's your story.

If so, let me encourage you.

God has never been surprised by your circumstances.

He knew this season was coming long before you did.

Sometimes He allows a closed door because He intends to open a different one.

Sometimes what feels like an ending is actually a new assignment.

God never wastes experience.

Your greatest ministry may begin after your greatest disappointment.

Your scars may become someone else's survival guide.

Perhaps this isn't the end.

Perhaps this is your pivot.

Fear Is the Enemy of Hope

Fear has one purpose.

It wants to convince you that tomorrow will be worse than today.

Fear asks,

"What if I fail?"

Faith asks,

"What if God keeps every promise He has ever made?"

Fear magnifies problems.

Faith magnifies God.

Scripture repeatedly says,

"Fear not."

Why?

Because fear causes us to forget who is walking with us.

Isaiah 41:10 reminds us:

"Fear not, for I am with you; be not dismayed, for I am your God. I will strengthen you. Yes, I will help you. I will uphold you with My righteous right hand" (NKJV).

That promise changes everything.

God Is With You

One of the greatest truths in Scripture is not merely that God exists.

It is that God is with us.

Think about that.

The Creator of heaven and earth . . .

THE ONE WHO SPOKE GALAXIES INTO EXISTENCE . . .
THE ONE WHO DEFEATED DEATH . . .
THE ONE WHO NEVER CHANGES . . .

He is with you.

Not just watching you.

Walking with you.

Providing for you.

Strengthening you.

Guiding you.

Throughout Scripture, God revealed Himself through names that describe His character.

Every one of those names answers one of our greatest fears.

When you fear lack, He is Jehovah Jireh, the Lord who provides. When you fear sickness, He is Jehovah Rapha, the Lord who heals. When anxiety overwhelms you, He is Jehovah Shalom, the Lord our peace. When you feel alone, He is Jehovah Shammah, the Lord who is there. In battles beyond your strength, He is Jehovah Nissi, the Lord our banner. When you need guidance, He is Jehovah Rohi, the Lord our Shepherd. When you struggle with guilt, He is Jehovah Tsidkenu, the Lord our righteousness. And as God shapes and transforms you, He is Jehovah Mekadesh, the Lord who sanctifies.

I wish I had space in this book to unpack each of these names more fully. I encourage you to identify the area in which you struggle most. Study that name of God, pray over it, and meditate on it until the truth of who He is becomes real in your heart. For an example, I had great fear around the Lord as my provider.

Learning That God Was My Provider

I learned this lesson as a young husband.

Sandra had made the decision to stay home and homeschool our children.

I believed that was exactly what God wanted for our family.

But I was terrified.

How was I going to provide?

I was young.

We had three children.

Money was tight.

On top of that, I carried fears that many black men have carried for generations. I wondered if opportunities would be available to me. I wondered if I would ever be able to provide the life I wanted for my family.

Fear had become louder than faith.

So I did something that changed my life.

I asked my pastor if I could lock myself in the church and stay there for three days.

For three days, I searched the Scriptures.

I wasn't looking for sermons.

I wasn't preparing a message.
I was looking for God.

I found every verse I could about God's provision.

I read them.

I prayed them.

I spoke them aloud.

Slowly, something began to change inside me.

The circumstances hadn't changed.

My income hadn't changed.

But my heart had changed.

Faith comes by hearing, and hearing by the Word of God.

As God's Word filled my heart, fear began to leave.

I stopped seeing myself as my family's ultimate provider.

I began seeing God as my provider.

Yes, I had the responsibility to work.

Yes, I had the responsibility to lead my family.

Yes, I even encountered racism from time to time. But, because my identity was in Christ and I am seated with Him above all principality and power (see Ephesians 1:19-23, 2:6-7), therefore no thing and no one will stop me.

Because Jehovah-Jireh was the One supplying every need.

Then one afternoon sometime later, something happened that I will never forget.

Our daughter Tasha came through the front door after playing outside.

She looked at me and asked,

"Daddy, how are we going to pay our bills?"

I smiled.

I said,

"Tasha, that's not your responsibility. Go back outside and play."

She simply turned around and walked outside.

Then the Holy Spirit spoke quietly to my heart.

"John . . .

Why are you worrying about your bills?

I'm your Father.

You go play."

I have never forgotten that moment.

God wasn't telling me to become irresponsible.

He was teaching me to trust.

There is a huge difference.

That day I realized something that has carried me through every season of my life.

I am not the source.

God is.

Learning that God is my provider changed everything. It became more than a verse—it became a revelation. Years later, that revelation was put to the test when I suddenly lost my income. By every human standard, I had every reason to panic. But I didn't. Why? Because Jehovah Jireh was already settled in my heart. I knew God was my provider—not my paycheck. He carried me through that season and faithfully brought me to the other side.

Now let me ask you: Which aspect of God's character do you need to know more deeply? Is it His provision? His healing? His peace? His presence? His guidance?

Don't wait until the storm comes to discover who God is. Study His Word. Pray. Meditate on His promises until His truth moves from information to revelation.

Faith and fear cannot coexist in your heart. As your faith grows, your fear diminishes. Build your faith now so that when God calls you to step forward one more time, fear will not hold you back.

Hope is the quiet confidence that God is not done. Faith gives substance to that hope. Fear tries to silence it. If hope is lost, faith has nothing to stand on. As Hebrews 11:1 reminds us, "*faith is the substance of things hoped for*" Hope is the fuel that keeps faith alive, and faith is what carries you through every season God has prepared for you.

Hebrews 6:19 reminds us, "*We have this hope as an anchor for the soul, firm and secure.*"(NIV).Hope anchors your soul. Faith gives that hope substance. Fear threatens to silence it, but it never gets the last word when your hope is anchored in Christ.

Guard your hope. Feed your faith. Refuse to give fear a place in your heart. Keep your hope anchored in Christ, choose faith, and finish well.

One More Time

Maybe you've run out of hope.

Maybe you've lost your job.

Maybe retirement didn't happen the way you planned.

Maybe you've buried someone you love.

Maybe you're wondering if your best years are behind you.

Let me tell you what I have learned.

As long as God is with you . . .

There is still hope.

As long as God is your Provider . . .

There is still hope.

As long as God still gives you breath in your lungs . . .

There is still purpose.

You are not finished.

Your story is not over.

Your greatest contribution may still be ahead of you.

So get up.

Trust God again.

BELIEVE AGAIN.
DREAM AGAIN.
LOVE AGAIN.
LEAD AGAIN.

And with God's help . . . **Try one more time.**

Chapter 9 Application

Hope: The Fuel for One More Time

Reflect

Think about a time when you felt you had run out of hope. What changed when you chose to trust God's presence and His promises instead of your circumstances?

Examine

In what area of your life is fear causing you to believe that tomorrow will be no better than today? How would your perspective change if you truly believed God was already at work?

Remember

Recall a time when God revealed Himself as your Provider, Healer, Protector, or Shepherd. How does remembering His faithfulness strengthen your confidence for what you're facing today?

Renew

Hope must be cultivated. What practical habits can you develop this week to renew your hope through Scripture, prayer, worship, or encouraging relationships?

Respond

Where might God be inviting you to make a change? What feels like an ending, that could actually be the beginning of a new assignment?

ACT

If you truly believed your greatest contribution is still ahead of you, what would you begin again? What one step of faith can you take this week to move toward God's purpose?

FINAL REFLECTION

Are you allowing your present circumstances to define your future, or are you allowing God's promises to define your expectations? What would change if you believed your "one more time" could become the beginning of God's next chapter for your life?

CHAPTER 10

Do You Have Scar Tissue? Then You Are Qualified!

"From now on, let no one cause me trouble, for I bear on my body the marks of Jesus" (Galatians 6:17, NIV).

I once had a senior man tell me that he wasn't sure he was qualified to mentor younger men.

He said, "John, I've made too many mistakes. I don't know that I have anything to offer."

I asked him a simple question.

"Do you have any scar tissue?"

He paused.

"Yes," he said. "I do."

I replied, "Then you're qualified."

If you have scars, you have stories. And if you have stories, you have wisdom that the next generation desperately needs.

Every scar tells a story.

Some scars came from circumstances beyond your control. Others came from wrong turns, poor decisions, and moments when you knew you should have gone right, but chose left instead. Yet somehow, God intervened. He rerouted your journey.

Have you ever been driving with GPS and missed your turn? You hear that familiar phrase:

"Recalculating."

The destination doesn't change. The route does.

God does that with our lives.

He doesn't excuse our mistakes, but neither does He waste them. He teaches through them. He redeems them. He transforms them into testimonies.

Young people need to hear those stories.

In fact, they want to hear them.

They don't need mentors who pretend to have lived flawless lives. They need mentors who have wrestled with failure, experienced God's grace, and discovered wisdom through the journey.

As I reflect on the life of Samson, I see a man many of us can relate to.

Judges 16 records Samson's final prayer:

"O Lord God, [earnestly] remember me, I pray You, and strengthen me, I pray You, only this once, O God" (Judges 16:28b, AMPC).

Those words have shaped my thinking:

One More Time

Samson had a divine calling.

Before he was born, an angel announced his purpose. God had set him apart for a specific assignment. In the same way, every person has a God-given purpose.

Ephesians 2:10 reminds us:

"For we are God's masterpiece. He has created us anew in Christ Jesus, so we can do the good things he planned for us long ago" (NLT).

God has prepared paths for us to walk in. He has

purposes designed specifically for our lives.
Yet Samson lived inconsistently with that calling.

He made poor choices.

He violated his vows.

He pursued unhealthy relationships.

His life was messy.

And still, God used him.

In fact, Samson appears in Hebrews 11 among the great heroes of faith.

Think about that.

Abraham is there.

Moses is there.

Gideon is there.

And Samson is there too.

Messy Samson.

Broken Samson.

Failure-prone Samson.

Apparently, God has a different standard for usefulness than we do.

Perhaps your life hasn't unfolded the way you imagined.

Maybe you've made mistakes.

Maybe you've squandered opportunities.

Maybe you've experienced divorce, addiction, financial collapse, moral failure, or broken relationships.

Maybe you've looked back and thought, "I should have accomplished more by now."

But what if the very things you regret most have become part of the story God intends to use?

We often assume our greatest impact comes through our victories.

I don't think that's true.

Our deepest impact often comes through our failures.

Through the lessons learned in pain.

Through the humility gained in weakness.

Through the wisdom acquired by surviving what once threatened to destroy us.

The scars become sacred.

Scripture repeatedly shows God using human stories—even painful ones—to teach future generations what to pursue and what to avoid.

There is a generation rising behind us.

They need wisdom.

They need access to experience.

They need people willing to say:

"I've walked this road. Let me show you where the potholes are."

I often tell young men, "You don't have to make every mistake yourself."

This happened early in our marriage, when we were still trying to navigate through life issues. It was 20 degrees F in the early 1980s in Cincinnati. Sandra and I argued about something. I got so mad and decided to go outside late that night. I remember slamming the door as hard as I could, when I went outside. You know, I tried to shake the apartment, so she could know how pissed I was! I sat in my 1973 Cutlass with a hole in the floor and a bad muffler. I started the car up to stay warm, but the fumes from the muffler came up through the floor. I thought, "If I keep the car running, I will probably die in here!" So, I turned the car off.

As I sat in the very cold car, the Scripture came to mind, *"Be angry and do not sin; do not let the sun go down on your anger"* (Ephesians 4:26, ESV).

No doubt the Holy Spirit was trying to help me. I remember saying, "Yeah, I know that verse, but it says don't go to bed angry. So, I will stay up all night; this way I am not disobeying the Word, and I can still be pissed! But that's not what it says! *"Be angry and do not sin; do not let the sun go down on your anger"* (Ephesians 4:26, ESV). What is it saying here? Resolve your issues, and don't carry them over to the next day, and the next day, and the next day!

It was getting later and later (probably 3:00 in the morning). I was wrapped up in a blanket, shivering! I remembered we had a very nice warm water bed. We bought it so we could have great sex! I begin to think, "Here I am outside in the cold; I will probably freeze to death; and I have a gorgeous woman upstairs in that nice warm water bed! For what? This is really stupid!" So, I sheepishly went back into the house, only to find Sandra asleep! I woke her up and apologized. "I am sorry, Sandra."

Now, when I share that story with young husbands, I tell them:

"You don't have to do that. I've already done it for you."

Wisdom allows others to learn through our experience, rather than through unnecessary pain.

They can skip over some of the mistakes we made.

Our mistakes can accelerate their growth.

That's the gift of mentoring.

Your scars become road signs.

Your failures become warnings.

Your restoration becomes evidence of God's grace.

So, if you're wondering whether you're qualified to invest in the next generation, let me ask you the same question I asked that senior man.

Do you have scar tissue?

Have you known disappointment?

Have you experienced heartbreak?

Have you made bad decisions and lived long enough to learn from them?

Have you fallen and gotten back up?

Then you have something valuable to offer.

Because every scar has a story.

And every story, surrendered to God, becomes a source of wisdom for those who follow behind.

Perhaps your greatest contribution isn't found in your successes.

Perhaps it lies in what God taught you through your failures.

Maybe your prayer, like Samson's, is simply this: "Lord, strengthen me just one more time."

One more time to encourage.

One more time to teach.

One more time to invest.

One more time to help someone else avoid unnecessary pain.

One more time to fulfill the purpose for which You created me.

Your story isn't over.

Your scars still speak.

And someone needs to hear what they have to say.

Scar Tissue Becomes Your Brand

In the ancient world, slaves were often branded with the mark of their master. Soldiers sometimes bore marks identifying their allegiance. Those marks answered one question:

Who do you belong to?

Paul looked at the scars across his body and said, *"I bear in my body the marks of the Lord Jesus"* (Galatians 6:17, NKJV).

He wasn't ashamed of those scars.

He wasn't trying to hide them.

He wasn't trying to explain them away.

Those scars proved that he belonged to Christ.

The false teachers boasted about circumcision.

Paul boasted about scar tissue.

One pointed to religious performance.

The other pointed to faithful perseverance.

I think many Christians spend their lives trying to hide their scars.

We hide our failures.

We hide our addictions.

We hide our divorces.

We hide our fears.

We hide the years we wandered away from God.

We hide our disappointments.

But God often says,

"That's the very thing I want to use."

Every scar has a story.

Every story has a lesson.

Every lesson has the power to help someone else.

Your scar tissue becomes your ministry.

Your weakness becomes your witness.

Your pain becomes your platform.

I remember something that someone asked me, after I shared some painful stories from my marriage and from mistakes I'd made as a husband.

He said,

"Aren't you embarrassed to tell people those stories?"

I smiled and said,

"If my goal was to impress people, then yes, I'd be embarrassed."

"But, I'm not trying to impress people."

"I'm trying to impact people."

I've learned something over the years.

WE IMPRESS PEOPLE WITH OUR STRENGTHS. WE IMPACT PEOPLE WITH OUR SCARS.

People admire success.

But, they identify with struggle.

No one needs another hero pretending they've never failed.

People need someone who can honestly say,

"I've been there."

"I know what failure feels like."

"I know what regret feels like."

"I know what forgiveness feels like."
"And Jesus met me there."

So don't be ashamed of your scar tissue.

Don't hide what God has healed.

Those scars are no longer evidence of your defeat.

They are evidence of God's grace.

They are your brand.

They declare,

"I belong to Jesus."

And because of that . . .

You're qualified.

Qualified to encourage.

Qualified to disciple.

Qualified to mentor.

Qualified to tell your story.

Not because you've lived a perfect life . . .

but because you've experienced a perfect Savior.

The Apostle Paul writes:

"These things happened to them as examples and were written for our instruction . . ." (1 Corinthians 10:11, NET Bible).

God never intended biblical stories to be admired merely as history. They were preserved because they reveal how God works in every generation. The stories become invitations for us to trust Him in our own circumstances.

One of those stories is the woman with the issue of blood.

For twelve long years, she suffered. She had exhausted every resource. Mark tells us she had spent all she had on physicians and was no better—in fact, she had grown worse. Humanly speaking, she had run out of options.

Then she heard about Jesus.

Faith often begins with hearing someone else's story.

She said within herself,

"If only I may touch His clothes, I shall be made well" (Mark 5:28, NKJV).

She pushed through the crowd, knowing the tremendous risk she was taking.

According to the Law, her continual bleeding made her ceremonially unclean. Anything she touched became unclean. Anyone she touched became unclean. By entering that crowd, she risked public humiliation and severe consequences. She certainly didn't want anyone to know what she had done.

The moment she touched Jesus's garment, the Bible says she **felt in her body that she was healed.** Immediately, the flow of blood stopped.

Her first instinct was not to testify.

Her first instinct was to disappear.

But Jesus wouldn't let her leave anonymously.

He asked, *"Who touched My clothes?"* (Mark 5:30, NKJV).

Not because He lacked information.

He wanted to give her something greater than physical healing.

He wanted to give her a testimony.

When she realized she could not remain hidden, Mark says:

"The woman, fearing and trembling, knowing what had happened to her, came and fell down before Him and told Him the whole truth" (Mark 5:33, NKJV).

She confessed everything.

Her pain.

Her desperation.

Her faith.

Her healing.

Then Jesus spoke words that forever changed her identity:

"Daughter, your faith has made you well. Go in peace, and be healed of your affliction" (Mark 5:34, NKJV).

Notice that Jesus did not shame her.

He honored her.

He transformed what she wanted to hide into something the world would remember for thousands of years.

Her testimony produced faith in others.

Later in Mark's Gospel we read:

"Wherever he entered . . . they laid the sick in the marketplaces and begged him that they might just touch the fringe of his garment; and as many as touched him were made well" (Mark 6:56, WEB).

Why were they asking to touch the fringe of His garment?

Someone had to tell them.

The woman who once wanted to disappear became part of the story that spread throughout Israel.

People heard what Jesus had done for her.

If He healed her . . .

Maybe He could heal me.

Her testimony became their hope.

That is why God doesn't waste your scars.

The enemy wants you to hide them.

Jesus wants you to redeem them.

The enemy whispers, "If people knew what you've been through, they would reject you."

Jesus says, "Tell them the whole truth."

Because your story may become someone else's breakthrough.

Your scars may become someone else's faith.

Your failure may become someone else's warning.

Your restoration may become someone else's hope.

Paul says these things were *"written for our instruction"* (Romans 15:4, ESV). Every biblical testimony teaches us something about the character of God. And every modern testimony does the same.

Don't hide what God has healed.

Don't cover what God has restored.

Don't be ashamed of the marks you've received while following Christ.

We learned in Galatians 6:17, Paul didn't hide his scars—he called them ***"the marks of Jesus"*** (NIV). The woman didn't remain anonymous—**she told Him the whole truth.** In both cases, what could have been a source of shame became a witness to God's grace.

Someone today is waiting to hear your story.

Because when they hear what Jesus has done for you, they may finally believe that He can do it for them.

Your greatest qualification for ministry is not the absence of scars—**it is the presence of God's grace in the middle of them.**

Chapter 10 Application

Do You Have Scar Tissue? Then You Are Qualified!

Reflect

Think about a painful season, failure, or disappointment that has left scars in your life. What has God taught you through that experience, that you couldn't have learned any other way?

Examine

Is there a part of your story you've been reluctant to share because it brings back feelings of shame, regret, or embarrassment? What would it look like to trust God, by allowing that scar to become a testimony of His grace?

Remember

Looking back, where have you seen God "recalculate" your life? How has He redeemed a wrong turn, restored what was broken, or used your mistakes for His purposes?

Invest

Who in your life could benefit from the wisdom you've gained through your struggles? What lesson have you learned that could help someone else avoid unnecessary pain?

Encourage

I write that "your scars become road signs." If a younger believer asked for your advice today, what story from your own life would you share to encourage and strengthen them?

Pray

Ask God to replace any lingering shame with gratitude for His redeeming grace. Pray for the courage to tell your story honestly, trusting that He can use your scars to bring healing and hope to others.

Act

This week, intentionally share one story of God's faithfulness with someone who needs encouragement. Don't focus on impressing them with your successes. Point them to the grace of God that met you in your weakness.

Final Reflection

What if your greatest qualification for serving others isn't your success—but the grace of God displayed through your scars? What testimony has God entrusted to you, that someone else needs to hear today?

CHAPTER 11

LETTING GO TO TAKE HOLD!

Sometimes God Asks You to Leave What You Love

Martial arts had been a tremendous blessing in my life. It taught me discipline, perseverance, confidence, leadership, and the importance of continual growth. Through martial arts, I had the opportunity to train students, build relationships, and watch people develop both physically and mentally. It was much more than a hobby—it was part of who I was.

At that particular season of my life, I was preparing for my next level of training. I had become fascinated with another martial arts discipline—judo. I absolutely loved it. I enrolled in a school and began learning everything I could.

I have always had a bit of a playful side, so

whenever I came home from training, I wanted to practice what I had learned on my family. My wife, Sandra, and the children quickly learned what that meant. Sometimes when I walked through the door, they would dramatically fall to the floor before I could get near them, because they knew I was looking for someone to throw. We laughed about it often, and those memories still make me smile today.

At the same time, my martial arts school was thriving. We had developed several outstanding students. Some were champions who competed successfully in tournaments. Others were growing in confidence and character through their training. I loved teaching them. I loved watching them mature.

Sandra was right there beside me, just as she has been through so many seasons of my life. Whenever God placed something in my heart, she was always willing to serve alongside me. She became my assistant instructor. My daughter Rhea was also progressing through the ranks and had reached the brown belt level with a red tip. Even my younger children were involved in the school.

Life was good.

The school was growing.

The students were thriving.

And I was doing something I loved.

Then God spoke.

I remember it vividly. It was a Saturday. I had been praying, seeking the Lord, and asking Him to direct my life.

"Lord, I want whatever You want for me."

As I prayed, I sensed the Spirit of God speak clearly to my heart.

"Drop the judo."

The words hit me unexpectedly.

Drop the judo?

I didn't understand. I was enjoying it. I was learning. I was growing.

Then came the second instruction.

"You need to continue in what I have already given you. This is a distraction from the next thing I have for you."

That was difficult enough.

Then came the third instruction.

"Close the martial arts school."

My heart sank.

I felt as though time stopped.

Close the school?

Martial arts wasn't just something I did. It was a major part of my life. It represented years of investment, countless hours of training, friendships, accomplishments, and dreams.

It felt like God was asking me to place something precious on the altar.

Immediately, my mind went to Abraham on Mount Moriah.

When God told Abraham to sacrifice Isaac, He wasn't asking for something Abraham didn't care about. He was asking for the thing Abraham loved most. Isaac represented the promise. Isaac represented the future.

In my own small way, I felt like I was standing on a similar mountain.

God was asking me to trust Him enough to surrender something I loved.

One of the great lessons of life is that whenever God wants to take you into a new season, He will often ask you to release something from the old season.

You cannot always carry yesterday's assignment into tomorrow's calling.

**SOMETIMES IT'S A CAREER.
SOMETIMES IT'S A POSITION.
SOMETIMES IT'S A RELATIONSHIP.
SOMETIMES IT'S A DREAM.**

Sometimes it's simply a comfortable place that you've outgrown.

Many people never reach their next level because they refuse to let go of their current one.

But faith requires open hands.

Even today, I understand that lesson better than ever. There are things in every season that God asks us to surrender. Letting go is never easy. In fact, it can be painful. Yet, surrender is one of the greatest expressions of trust.

Eventually, I shared what God had spoken, with Sandra and my family. Then came the hardest part.

I had to tell the students.

Some of them were close to earning their black belts. I had dreamed of standing beside them and watching them reach that milestone. Their parents had invested years into their training. We had built a community together.

The day I announced the school's closing was heartbreaking.

The students were shocked.

The parents were shocked.

There were tears.

There were questions.

There was disappointment.

There was also tremendous appreciation for what we had built together.

As I stood before them, my heart was broken. Yet deep inside, I knew obedience was more important than comfort.

God was leading me somewhere I could not yet see.

Just as He told Abraham, *"Leave your country, and your relatives, and your father's house and go to the land that I will show you"* (Genesis 12:1, WEB), He was asking me to leave a familiar place and trust Him for the next chapter.

That's the challenge of faith.

God rarely shows us the entire journey.

He simply asks us to trust Him for the next step.

Looking back now, I can see that what felt like a loss was actually preparation. God wasn't taking something from me; He was positioning me for something greater. What seemed like an ending was actually a beginning.

Sometimes the road to God's next assignment runs directly through the altar of surrender.

And every time we place something in God's hands, we discover that His plans are always better than our own.

As difficult as it was, I knew it was time to let it go and move forward. So I laid it aside and stepped into the next season of my journey, trusting that God was continuing to shape my life according to His purpose.

Sometimes you have to let something die for you to live again.

When I was fifty-nine years old, life looked successful by most standards. I had been a pastor for more than thirty years. I had started businesses, including a digital marketing company. I served in leadership roles in several ministerial associations. I had launched a nonprofit organization designed to help pastors and churches. On the surface, everything seemed to be moving in the right direction.And to be honest, I enjoyed it.

I loved building. I loved leading. I loved helping people. I had developed skills over the years and had become proficient in many different areas. I wasn't content to do just one thing, when I believed I could do several.

But beneath all the activity, something was stirring inside me.

I knew I couldn't sustain the pace forever.

Then one day, a friend asked me a question that changed my life.

"John," he said, "you're doing a lot of things. If you could only do one thing, what would that be?"

His words caught me off guard.

I had always believed that being productive was a virtue. Doing more meant accomplishing more. But his question forced me to confront something I had carefully avoided: just because I could do something didn't mean *I should.*

Over the years, I had taken the talents God had given me and developed them. I often think of myself as someone who started out as a one- or two-talent person, who simply tried to be faithful with what he had. And in His kindness, the Lord multiplied those talents.

But there is a danger in becoming multi-talented.

The more capable you become, the more opportunities come your way. Eventually, you can find yourself doing everything simply because you can.

That was exactly where I found myself.

Around that same time, my wife, Sandra, and I were celebrating our anniversary. For more than twenty-five years, we had made it a practice to spend time praying together, seeking God concerning our marriage and the direction of our lives. We developed annual marriage plans and asked the Lord how we could grow closer as a couple and better reflect His purposes.

During one of those times away together, I asked God a simple, but life-altering question.

"Lord, what should I be doing in this season of my life?"

I was fifty-nine years old.

I quieted my heart before Him and waited.

The answer came clearly.

"I want you to pour yourself out to the next generation. Anything that is not focused in this area, either adjust it so that it becomes the focus, or release it." I knew immediately that what I had heard would require significant change.

What I didn't know was how painful that change would be.

Over the next two years, God began leading me through a process of surrender.

Some things were easy to release.

Others felt like death.

At the center of it all was the church I had planted.

It was my baby.

I had dreams for that church. I had poured years of prayer, sacrifice, and effort into it. I had watched people come to Christ. I had witnessed transformed lives. Yet, if I were honest, I also knew that we had fallen short of the vision God had originally placed in my heart decades earlier.

I had learned how to develop individuals.

What I had not fully learned was how to build and lead teams.

As I attempted to reposition the church toward reaching the next generation, I encountered resistance. I tried to adjust. I tried to recruit new leaders. I tried to inspire those already around me. But nothing worked.

It seemed as if the grace, that had once rested

upon that assignment, had lifted.

People would say to me, "John, we love you. We believe in you. We want to work with you. But we don't want to work in your church."

Those conversations wounded me.

I didn't understand them.

I took them personally.

Yet, over time, I realized they were revealing something I had been unwilling to acknowledge.

God was doing something new.

After much prayer, fasting, and conversations with Sandra, I sensed the Lord speaking to me again.

"Why are you holding on to this? This is no longer what I am calling you to do. You need to let it die." I struggled with those words.

After all, I had spent years preaching, "Never quit. Keep going. Finish strong."

One of my favorite statements that I would make especially during my sermons—to encourage people to keep believing, to keep enduring hardness like a soldier—was, "It's my ball, my bat, my backyard, my game. I play till I win!" To let my church die was quitting in my mind!

But I was learning that sometimes faithfulness means having the courage to release what God is no longer blessing.

Eventually, I shared the decision with my children. Some cried.

Others were upset.

One asked, "Are you going to abandon your vision?" That cut me deep!

"What's next? Are you and Mom going to tell us you're getting a divorce?"

I found it interesting, that even those who were no longer actively involved, still felt a profound sense of loss.

Then came the conversations with our church family.

We held a final service.

It felt like a funeral.

There were tears.

There was grief.

There were questions.

Then something happened that I will never forget.

Several members gathered around me and said, "Pastor, we release you. We believe God has something else for you to do. We cannot hold on to you simply because it makes us comfortable. We release you to fulfill God's purpose."

As they prayed over me, I wept.

It was one of the most painful—and one of the most sacred—moments of my life.

That season taught me something I had never fully understood before.

For years, I had invited God to become a partner in my plans.

What I eventually discovered was that God had no interest in partnering with my agenda.

Instead, He was inviting me to join Him in His.

I began to notice that God was already at work.

He was moving in communities.

He was moving in families.

He was moving in the hearts of emerging leaders.

And once again, I heard His invitation:

"Pour yourself out to the next generation."

As I aligned my life with what God was already doing, everything changed.

Doors opened.

Relationships flourished.

Opportunities multiplied.

For the first time in years, I experienced the joy that comes from living in step with God's timing and purpose.

I have come to believe that the final quarter of life is not a season of retreat.

It is a season of significance.

The younger generation does not merely need our opinions.

They need our understanding.

Knowledge is information.

Wisdom is applied knowledge.

But understanding is wisdom that has been tested over decades and proven faithful.

That is what we possess.

We have scars.

We have stories.

We have victories.

We have failures redeemed by grace.

And now, perhaps more than ever before, the world needs what God has been forming within us all along.

This is not the time to withdraw.

This is the time to pour ourselves out.

This is our hour.

We have come to the Kingdom for such a time as this.

The Armor I Was Never Meant to Wear

Sometimes you have to let go of the pain, let go of the hurt, even when you think you already have, because forgiveness and healing come in layers.

When I first began pastoring, I knew God had called me. There was an anointing on my life. People were coming to Christ. Marriages were being restored. People were healed, delivered, and discipled. Week after week, I poured myself into preparing messages, praying for people, and serving with everything I had.

Every few years, something would happen that caused me to shut down in another area of my life.

I remember the first major thing that happened involved my own family. Some of my relatives were part of the church I was pastoring, and through that situation I experienced deep hurt and disappointment.

I remember saying in my heart, "I'll never let family get that close to me again. I'll never allow my own family to hurt me like this again."

I didn't say those words out loud.

In fact, I didn't even realize I had made that decision.

But I had.

A door quietly closed in my heart.

I continued preaching. I continued serving. I continued loving people. Outwardly, nothing seemed to change. Inwardly, however, I had begun protecting a part of my heart that I would never allow anyone—even my own family—to touch again.

I thought I was protecting myself from future pain.

I didn't realize I had begun building armor instead of trusting God to heal my wounds.

Without realizing it, I made an inner vow:

"I'll never let family get that close to me again."

Life moved on.

One of my closest friends from Bible school moved to Atlanta to work with me in the ministry. We had made a covenant before God. We had promised to stand together wherever God led us. We had to address some issue he was having in his marriage. This caused him to walk out of my life and from the work we were doing together.

When he left, it felt like someone had driven a dagger into my heart.

Worse yet, he began speaking against me behind my back, trying to undermine my ministry.

Again I made another vow.

"I'll never let another minister get that close to me again."

Years passed.

Then church members whom I had loved, discipled, and sacrificed for, turned against me. Some tried to divide the church. Others attempted to destroy my reputation.

Another door quietly closed.

Then associate ministers.

Then pastors.

Each betrayal added another layer.

One particular experience shook me deeply.

One of the men involved—a white leader who had become an influential part of our church—came to me and said, "Pastor John, this is an amazing, loving church. In fact, I've never experienced love like this before. You are one of the best pastors I've ever had, and I have learned so much from your teaching. But I just can't get past the fact that you're black. I don't know how I can continue attending this church or invite my friends here."

Soon afterward, he left. One by one, nearly all of my white members left with him. Only one single white woman remained. I was devastated.

I had carried a God-given dream of building a Kingdom ministry where people from every race and background would worship, grow, and serve together as one family. In what felt like a moment, that dream died in my heart.

The pain was so deep that I found myself embracing something my mother had often said while I was growing up, "You can't trust white people."

Looking back, I realize I wasn't responding from faith—I was responding from woundedness. A painful experience had begun shaping my beliefs more than God's truth.

What one group of individuals had done, I unconsciously projected onto an entire race.

That wound would become one of the greatest barriers God would later ask me to confront and surrender.

I remember saying before the Lord,

"I'm never going to pour my life into white people again."

It wasn't hatred.

It was self-protection.

Looking back now, I realize I wasn't just withdrawing from certain people. I was withdrawing from anyone who had the potential to hurt me.

Every disappointment became another brick in a wall around my heart.

I thought I had forgiven everyone.

But forgiveness had never reached the deepest places.

Instead of putting on the armor of God, I had slowly forged my own armor.

It wasn't made of truth, righteousness, faith, or salvation.

It was made of disappointment.

BETRAYAL.
FEAR.
SELF-PROTECTION.

The frightening part was this:

I didn't even know I was wearing it.

Then one day, everything changed.

Sandra and I were standing together at the altar during a service, where the Holy Spirit was moving powerfully.

Suddenly I felt something happen inside me.

It was like I felt something pulled off of me, like a sweater pulled off and over my head.

Instantly, all the pain I had been protecting myself from, came rushing back.

In my mind, I could see wound after wound across my heart. They were open, infected, and bleeding.

Then it was as though a scroll began to unroll before me.

Face after face appeared.

Every person who had betrayed me.

Every person who had lied about me.

Every person who had rejected me.

Every person who had wounded me.

Every person who attacked my wife.

Every person who hurt children (and, some have still not recovered fully).

And with every face, came the exact pain I had felt the day it happened.

It wasn't old pain.

It felt brand-new.

Then the Lord spoke to my heart.

"Look at you."

"I can't use you like this."

"You have to let this go."

I was overwhelmed.

Sandra later told me that my face looked terrified.

She didn't know what I was seeing.

The only way I could describe it was that I felt spiritually naked. Everything I had hidden behind for years, was suddenly gone.

I collapsed to the floor.

I wept.

I sobbed before the Lord.

One by one, I forgave.

One by one, the Lord began healing wounds I didn't even know were still alive. As every offense was released, it was as if another piece of that false armor fell away. For the first time in years, my heart felt free. I wasn't protecting myself anymore—I was allowing God to protect me.

That day, I learned something I have never forgotten.

Sometimes you have to let go of offenses you thought you had already forgiven. Healing often comes in layers. God will gently uncover places in your heart you didn't even know were still wounded—not to shame you, but to heal you.

One Sunday during a special prayer conference at our church, a dear friend of mine—a white Bible teacher—was speaking. As they ministered, the Holy Spirit interrupted my thoughts and spoke clearly to my heart.

"John, you have abandoned the vision I gave you to build a multicultural Kingdom ministry."

I quietly responded, "Yes, Lord ... but, You know what happened."

With His familiar gentleness, He answered, "Yes, I know what happened. But that is no excuse for quitting on the call I have placed on your life. You will have to repent."

His words pierced my heart.

At the end of the service, I stood before my congregation and confessed my sin. I told them that I had allowed the pain of racism to become unforgiveness, and in doing so, I had abandoned God's vision for my life. I said, "Today I repent. I choose once again to believe God for a multicultural Kingdom ministry. I choose to believe in reaching white people again."

I could barely finish speaking. I broke down and sobbed.

What happened next is something I will never forget.

My congregation surrounded me. My dear friend stood with them. The group embraced me, and prayed over me, and said, "We love you, Pastor. We believe in you. And we will believe with you for the vision God has given you."

In that moment, something that had died inside me began to live again.

Forgiveness reopened the door that pain had closed. Repentance restored the vision that disappointment had buried.

Don't let yesterday's pain determine tomorrow's purpose. Repent where you need to repent. Forgive where you need to forgive. Pick up the vision God gave you. And trust Him . . . one more time.

Maybe that's where you are today.

Perhaps someone wounded you so deeply that you quietly laid down a dream God had placed in your heart. Maybe betrayal convinced you that it was safer not to believe again. If so, hear the gentle voice of the Holy Spirit as I did:

"Yes, I know what happened. But that is no excuse for abandoning My call on your life."

As you've been reading this chapter, faces have come to your mind. Names have surfaced. Old conversations have replayed. You have remembered betrayals, disappointments, broken promises, and deep hurts that you thought were buried.

Don't push them back down.

Bring them into the light.

Forgive.

Release them to God.

You may feel justified in holding on to the offense, but every offense you keep becomes another piece of armor around your heart. That armor may protect you from being hurt again, but it will also keep you from loving, trusting, and serving with your whole heart.

God cannot heal what we refuse to surrender.

The greatest enemy of my future was not the people who wounded me—it was the armor I built to make sure no one ever could again.

If He is going to strengthen you one more time, He must first heal you one more time.

Lay down the offense.

Take off the armor you built.

Put on the armor of God.

Then, with a healed heart, step back into His calling . . . one more time.

Chapter 11 Application

Letting Go to Take Hold!

Reflect

Is there something in your life—a role, a dream, a relationship, a position, or even an identity—that you sense God is asking you to release so He can prepare you for something new?

Examine

Think about a season when you held on to something longer than God intended. Looking back, what did that experience teach you about trust, surrender, and God's timing?

Heal

Are you carrying unresolved hurt, disappointment, betrayal, or bitterness that has caused you to build walls around your heart? What offense do you need to release, so God can continue His healing work in you?

Discern

Is there an area where you are asking God to bless your plans, instead of asking Him where He is already at work? How might your life change if you joined His agenda, instead of asking Him to join yours?

Trust

What "Isaac" has God placed on the altar in your life? How can surrendering what you love become an act of faith, that opens the door to His greater purpose?

INVEST

How could releasing yesterday's assignment free you to invest more intentionally in the people and purpose God has placed before you today?

PRAY

Ask God for open hands and a healed heart. Pray for the courage to surrender whatever He asks of you, the grace to forgive those who have wounded you, and the faith to embrace the next season He has prepared.

ACT

Write down one thing God may be asking you to let go of and one step you can take this week to move toward the new assignment He is placing before you.

FINAL REFLECTION

Sometimes the greatest obstacle to God's next assignment isn't the future—it's our unwillingness to release the past. What are you holding onto, that God is asking you to place in His hands, so you can fully step into His purpose?

What piece of self-protective armor have you been wearing, that God is asking you to remove? How might a healed heart allow you to love, trust, and serve more freely?

CHAPTER 12

When I Got Off God's Schedule

It was August of 2011, when I sensed a major shift in my life. I was fifty-two years old, and I knew a new season was about to begin.

Over the years, I had learned that God often spoke to me in decades. He would show me what the primary focus of each season of my life would be. When I was forty-five years old, the Lord spoke clearly to my heart.

"In your fifties, you will do three things: develop resources, develop leaders, and develop churches. Your pastoral ministry is coming to an end. Begin preparing for the next phase of your ministry."

So in 2011, I knew exactly what He meant. My assignment was changing.

As I began preparing my church for that transition, I came under one of the greatest spiritual attacks of my ministry.

A pastor, whom I had known for nearly twenty-five years, approached me with what seemed like a wonderful idea. He was an excellent Bible teacher, and he offered to establish a Bible school in our church. The plan was to train our leaders in theological studies and prepare them to become ordained elders and deacons. It sounded like exactly what we needed.

But Sandra had concerns.

She sensed something wasn't right about him.

Instead of listening to her, I shut her down.

Looking back, that was one of the greatest leadership mistakes of my life.

At that point, our church was healthy, growing, and filled with strong leaders, whom we had spent years developing. By opening our doors to this man, I unknowingly exposed those leaders to someone who had different motives.

What I believed would strengthen the church actually became the source of its greatest internal damage.

Once he began working with our people, he

became impressed with the quality of leaders God had raised up. Instead of helping develop them for the ministry God had given me, he wanted them for himself.

Like Absalom did with David, he quietly began undermining my leadership.

He questioned my decisions.

He discredited my authority.

He slowly attempted to steal the hearts of the people.

Had God not warned Sandra through a night vision, I might never have discovered what was happening. Through a series of events, the Lord exposed the entire plan, and I was forced to remove him from our church.

But by then, the damage had already been done.

It took several years to clean up the confusion and rebuild trust. Some relationships were permanently affected. Our momentum slowed, and the transition God had originally shown me was delayed.

One decision.

One ignored warning.

One season of misplaced trust.

It cost me years.

During that difficult season, while praying one day, the Lord spoke words that shook me.

"You are off schedule."

Those words pierced my heart.

The Lord continued,

"You are supposed to be developing resources, developing leaders, and developing churches. Instead, you are still pastoring. There are pastors and churches waiting for what I have placed inside of you, and you are not doing what I called you to do."

I answered,

"But Lord, remember what happened."

His response was immediate.

"Yes, I know what happened. That is why I have given you grace. But your time is up. If you are still pastoring in 2012, you will be outside of My will."

Those words filled me with holy fear.

Throughout my life, I had always lived by one personal conviction:

Never miss God on the major decisions of life.

You may miss Him on smaller things and recover.

But if you miss Him on the major assignments He gives you, the consequences can affect the rest of your life.

I had no desire to live outside the will of God.

Immediately, I gathered my leadership team and shared everything the Lord had spoken to me.

Together, we began preparing for the transition.

About that same time, I completed my certification with the John Maxwell Team and launched Preachers Trainer. Instead of focusing on building one local church, God expanded my assignment to equip pastors, develop leaders, create resources, and strengthen churches.

Looking back, I understand something I couldn't fully see then.

Sometimes, God allows painful delays to teach us lessons that success never could.

I learned to listen more carefully to the people God placed beside me—especially my wife.

I learned that discernment is just as important as vision.

And I learned that when God changes your assignment, obedience is more important than comfort.

Every season has an expiration date.

The blessing is not simply in being successful.

The blessing is in knowing when it is time to move.

Immediately after that meeting with my leadership team, we began putting a transition plan in place. It was not an easy decision. I loved my church, and I loved the people God had entrusted to my care. But I loved obedience even more.

Upon launching Preachers Trainer, I experienced something I had not felt in a very long time.

For the first time in years, I felt like I had stepped into the current of God's movement. The heavy lifting was over. Instead of constantly trying to force doors open, it seemed as though God was opening them for me. There was a grace on this assignment that I had not experienced before.

The ministry exploded.

We went from zero meetings to traveling nearly twenty days every month. Churches, from all over the country, began inviting us. We were equipping pastors, strengthening leaders, developing churches, and creating resources that were making a real difference. Everywhere we went, we saw leaders encouraged, churches strengthened, and pastors renewed in their calling.

I loved every minute of it.

As a pastor, I often found myself carrying the weight of making everything happen. I wore every hat. I solved every problem. I preached, led the staff, counseled families, managed finances, raised money, and tried to move the ministry forward. While God blessed our efforts, there were many seasons when I felt like I was pushing a heavy boulder uphill.

Preachers Trainer was different.

Instead of pushing, I found myself being carried by the grace of God. Opportunities came faster than I could have imagined. Relationships opened doors. Invitations multiplied. Resources flowed. The ministry continued to grow, and for the first time in a long time, I felt I was operating exactly where God had designed me to be.

There was a supernatural grace on my life.

Not only was the ministry flourishing, but financially we were doing better than we ever had. God was confirming that I had finally stepped into the assignment He had spoken to me about years earlier—to develop resources, develop leaders, and develop churches.

One of the greatest lessons I learned during that season is this: There is a difference between striving and grace.

When you are operating outside of your God-given assignment, everything feels heavier than it should. You can still accomplish good things, but you constantly feel as though you are fighting upstream.

But when you step into God's current assignment for your life, His grace begins to carry what you were never meant to carry alone. The work is still demanding, but there is a divine flow to it. You are no longer forcing open doors; you are simply walking through the doors that God has already opened.

Those years became one of the most fruitful seasons of my ministry. Looking back, I can clearly see that God was preparing me for something even bigger.

I didn't know it then, but another major transition was just around the corner.

When I turned fifty-nine, the Lord would once again speak to me about a new season—one that would stretch my faith, enlarge my capacity, and ultimately become the foundation for everything I am doing today.

Chapter 12 Application

When I Got Off God's Schedule

Evaluate

Looking at your current season of life, are you still serving yesterday's assignment, or have you embraced what God is calling you to do today? What evidence supports your answer?

Listen

God often confirms His direction through trusted people. Who are the "Sandras" in your life—people whose discernment you trust? Are you creating space to hear them, before making major decisions?

Identify

Is there an area of your life where you've confused faithfulness with staying too long? What might God be asking you to release, so you can embrace your next assignment?

Discern

When have you experienced God's grace carrying you, rather than your own effort sustaining you? What was different about that season?

Pray

Ask God to reveal whether there is any area where you are "off schedule." If He were to speak to you today about your current assignment, what do you think He might say?

CHAPTER 13

LET'S TALK ABOUT YOU

We've spent this book looking at Scripture.

We've looked at Joseph and betrayal.

We've watched Samson fall . . . and then rise one more time.

We've stood with David after failure.

We've followed Peter from condemnation to righteousness.

We've walked with the four lepers who asked,

"Why are we sitting here until we die?" (2 Kings 7:3, ESV).

We've talked about scar tissue.

We've talked about purpose.

We've talked about God's calling on your life.

We've talked about the Kingdom.

We've talked about hope.

But now . . .

I want to stop talking about them.

I want to talk about you.

Because somewhere inside every biblical story is your story.

Every one of us has experienced our own Egypt.

Our own wilderness.

Our own prison.

Our own cave.

Our own crossroad.

Joseph's story wasn't really about Joseph.

It was about betrayal.

Have you ever been betrayed?

Perhaps it was by a friend.

A spouse.

A business partner.

A pastor.

A church.

A family member.

Or someone you trusted with your heart.

Maybe you've experienced failure.

Not a small mistake.

A failure so public, so embarrassing, that you wondered if you would ever recover.

Maybe you've compromised.

Maybe you've sinned against God.

Maybe you've disappointed people who believed in you.

Maybe you've disappointed yourself.

Maybe you've carried guilt for years.

Maybe you've simply become tired.

Burned out.

Empty.

You kept serving.

You kept smiling.

But inside, you were exhausted.

I know something about those seasons.

There were times in my own life, when discouragement became so overwhelming, that I simply wanted to hide.

I remember lying in a dark room in the basement of my house with the covers pulled over my head.

Not wanting to answer the phone.

Not wanting to face another day.

Not because I had lost my faith.

But because I had lost my strength.

Perhaps you've been there too.

The longer I have lived, the more I've realized that life moves in seasons.

Winter.

Spring.

Summer.

Fall.

Winter is the season when everything appears dead.

Nothing seems to grow.

God has given you promises, but they seem buried beneath frozen ground.

Spring is the season of beginnings.

Small signs of life begin to appear.

Hope starts returning.

Summer is often the hardest season.

Growth requires heat.

Pressure.

Conflict.

Testing.

Discipline.

Much of what God develops in us happens during summer.

Then comes fall.

The harvest.

The season when everything God has been cultivating finally becomes visible.

The mature believer understands that every season has a purpose.

The goal is not to escape winter.

The goal is to remain faithful through every season.

Maturity is learning to maintain consistent rhythms regardless of what season you're in.

Prayer.

God's Word.

Fellowship.

Worship.

Obedience.

Faithfulness.

Those rhythms become the anchor of your soul.

So, here's what I'd like you to do.

Don't rush past this chapter.

Get a notebook.

Take out a blank sheet of paper.

Spend time with the Lord.

Write your story.

Where have you experienced betrayal?

Failure?

Disappointment?

Loss?

Burnout?

What dreams has God given you?

What dreams have died?

What prayers are still unanswered?

What scars has God redeemed?

Where have you seen His faithfulness?

Where has His grace carried you?

As you've read this book, what has the Holy Spirit been saying to you?

What keeps coming back to your mind?

What invitation is God extending to you?

Go back to the beginning.

Go back to the moment God first spoke.

Remember where the fire started.

Return to the place where you first heard His voice.

Every restoration begins there.

It begins with prayer.

It begins with fellowship.

It begins with sitting quietly before God, until His voice becomes louder than your fears.

Think about Samson.

By every outward measure, his life appeared finished.

His eyes were gone.

His strength was gone.

His ministry was gone.

His reputation was gone.

He was grinding grain in a prison.

Alone.

Broken.

Forgotten.

But Scripture quietly records one of the most hope-filled sentences in the Bible:

"However, the hair of his head began to grow again" (Judges 16:22, NKJV).

Perhaps your hair is beginning to grow again.

Perhaps your strength is returning.

Perhaps your faith is returning.

Perhaps your vision is returning.

Perhaps this book has simply been God's way of reminding you that your story is not over.

Isaiah paints another beautiful picture.

Israel had been reduced to a stump.

Everything appeared cut down.

Finished.

Over.

Yet God declared:

"A shoot will come up from the stump of Jesse; from his roots a Branch will bear fruit" (Isaiah 11:1, NIV).

Maybe today, you feel like a stump.

But God specializes in bringing life out of what appears dead.

He sends the roots deeper.

He nourishes them with living water.

He restores what was lost.

He strengthens what remains.

He causes life to emerge, where everyone else sees only endings.

The Lord is not finished with you.

This is not the end of your story.

It may simply be the beginning of your greatest season.

Hope is the anchor of the soul.

When Abraham could no longer see his future, God said,

"Lift up your eyes" (Genesis 13:14, ESV).

Look north.

South.

East.

West.

Everything *"you see, I will give to you"* (Genesis 13:15, ESV).

Later, God took him outside again and said,

"Look now toward heaven, and count the stars" (Genesis 15:5, NKJV).

As Abraham looked upward, faith returned.

Hope returned.

Vision returned.

Perhaps that's exactly what you need to do.

Lift up your eyes.

THE LORD WILL RESTORE YOU.
THE LORD WILL STRENGTHEN YOU.
THE LORD WILL HEAL YOU.
THE LORD WILL RAISE YOU AGAIN.

The Lord will fulfill every purpose He has ordained for your life.

One more time.

Chapter 13 Application
LET'S TALK ABOUT YOU

Reflect
As you look back over your life—through the victories, failures, disappointments, and restorations—where do you now see God's hand at work, even in the seasons when you couldn't recognize it?

Examine
What season are you in today? Is it a season of planting, growing, waiting, harvesting, or preparing for something new? How is God inviting you to remain faithful in this season?

Remember
Looking back over this book, what truth has impacted you the most? What lesson do you believe God wanted you to hear personally?

Listen
What has the Holy Spirit been whispering to your heart, as you've read these chapters? Is there an area of obedience, surrender, forgiveness, or faith that He continues to bring to your attention?

Respond

If you truly believed your greatest Kingdom impact could still be ahead of you, what would you do differently, beginning today? What dream would you revive? What person would you invest in? What step of faith would you finally take?

Commit

What one decision are you making because of what God has spoken to you through this book? Write it down. Share it with someone you trust. Then begin taking the first step toward living it out.

Pray

Father, thank You for reminding me that my story is not over. Thank You for Your faithfulness through every season of my life. Help me to release the past, embrace the present, and step confidently into the future You have prepared for me. Strengthen me one more time—not for my own success, but so I can glorify You, serve others, and leave a legacy of faith for those who follow. In Jesus's name, Amen.

Final Reflection

If this truly is your "one more time," how do you want the next chapter of your life to be written? What will people remember—not simply about what you accomplished, but about how faithfully you loved God, served others, and invested in the next generation?

CHAPTER 14

Four Pictures of Kingdom Succession

You can live your life one of two ways. You can pursue your own agenda and ask God to bless it, or you can discover what God is doing and hook your wagon to His.

The older I get, the more I realize that life is not primarily about accomplishing my plans. It is about participating in God's purpose.

The Apostle Paul wrote that we are *"his workmanship, created in Christ Jesus for good works, which God prepared beforehand that we should walk in them"* (Ephesians 2:10, ESV). There is a path God has already designed. There is a life He has prepared. Our responsibility is not to invent meaning but to discover it.

As we look back over our lives, we see victories and defeats. We remember dreams fulfilled and dreams that never materialized. We recall moments of great success and seasons marked by disappointment. Some memories bring gratitude; others bring regret.

Many older men begin asking difficult questions:

Did my life matter?

Did I miss my opportunity?

Is God finished with me?

The answer is simple:

If you're still here, God is not done with you.

You have survived hardships. You have endured losses. You have navigated seasons of uncertainty. You may carry physical limitations now. Your body may not respond as it once did. Your energy may not be what it used to be.

But if God has given you another day, then He has given you another assignment.

The question is not whether you still have purpose.

The question is whether you understand what that purpose looks like in this season of life.

The psalmist declared that the righteous *"shall still bear fruit in old age"* (Psalm 92:14, NKJV).

Fruitfulness does not expire. Purpose does not retire.

However, fruitfulness often changes form.

God may not call you to do what you once did, but He will always call you to become who He needs you to be.

I believe God gives us four powerful pictures to help us understand this season.

Picture One: Passing the Baton

The first picture is that of a relay race.

You have run your leg of the race.

There was a time when you were the one sprinting down the track. You built businesses. Led ministries. Raised children. Solved problems. Opened doors. Took risks.

And perhaps, by God's grace, you won a few races.

But there comes a moment when wisdom recognizes that the race is bigger than one runner.

In a relay, victory depends on the transfer.

The baton cannot remain in your hand forever.

Some older men spend their final years trying to outrun younger men. They cling to positions, titles, and recognition. They refuse to release what God entrusted to them.

But mature leadership understands something deeper:

Success is not finishing first. Success is ensuring someone else can continue the race.

There are young men in your life who need what you've learned.

**THEY NEED YOUR STORIES.
THEY NEED YOUR SCARS.
THEY NEED YOUR WISDOM.**

They need someone to tell them:

"Don't make the mistake I made."

"Here's what worked."

"Here's how God redirected me when I took a wrong turn."

You are no longer merely the runner.

You have become the coach.

The mentor.

The spiritual father.

The one entrusted with the baton.

Passing it requires humility.

It means celebrating another man's success.

It means finding joy in watching someone else run farther than you ever did.

John the Baptist modeled this spirit, when he said concerning Jesus:

"He must increase, but I must decrease" (John 3:30, ESV).

The baton was never ours to keep.

It was always ours to pass.

Picture Two: Stabilizing the Ladder

The second picture is a ladder.

Throughout life, we've been taught to climb.

Climb the corporate ladder.

Climb the ministry ladder.

Climb the ladder of success.

Climbing requires ambition, energy, resilience, and courage.

You understand what it means to climb because you've done it.

You've leaned ladders against the wrong walls.

You've fallen.

You've gotten back up.

You've discovered that some promotions cost more than they are worth.

But what happens when your season of climbing changes?

You become the one who stabilizes the ladder.

Anyone who has ever climbed understands the value of someone standing below.

Holding.

Supporting.

Protecting.

Spotting danger.

Offering encouragement.

You may no longer be the one reaching new heights.

But you can become the reason someone else reaches theirs safely.

You know where the ladder should lean.

You know which opportunities aren't worth pursuing.

You know the importance of character over charisma.

And so your success changes definition.

You move from the ladder of success to the ladder of significance.

You discover that significance is found in helping others rise.

Jesus said:

"The Son of Man came not to be served, but to serve" (Matthew 20:28, NIV).

That's your role now.

To steady.

To encourage.

To strengthen.

To hold the ladder.

Picture Three: Backstage Preparation and Support

The third picture is a stage.

There was a time when you were center stage.

You were "THAT" guy.

People applauded your accomplishments.

Your expertise was sought after.

Your performance mattered.

But seasons change.

Actors age.

Athletes retire.

Leaders transition.

Trying to remain the star, long after your season has passed, often diminishes the very legacy you hoped to preserve.

Instead, God may invite you backstage.

Backstage is where preparation happens.

It's where coaching happens.

It's where wisdom is transferred.

From backstage, you can tell younger performers:

"Here's how to handle criticism."

"Here's how to stay grounded during success."

"Here's how to navigate disappointment."

You also possess something many younger leaders lack: relationships.

You know people.

You have credibility.

You can open doors others cannot.

A phone call from you may create opportunities that would take someone else years to establish.

You become a connector.

A sponsor.

An advocate.

Your fulfillment comes not from hearing your own name announced, but from hearing someone else say:

"I wouldn't be here without the investment of this man."

The spotlight shifts.

But the impact increases.

Picture Four: Creating the Power Corner

The final picture is a boxing ring.

You remember what it meant to fight.

You trained.

You sacrificed.

You endured pain.

You experienced victories.

You suffered defeats.

You learned resilience.

But every boxer eventually reaches a point, where stepping back into the ring becomes dangerous.

Some fighters stay too long.

Pride convinces them that they still have something to prove.

Yet, wisdom recognizes that there is another role just as important.

The corner man.

The power corner.

The corner sees what the fighter cannot see.

They notice blind spots.

They identify openings.

They calm fears.

They treat wounds.

They provide perspective.

Between rounds they say:

"Keep your hands up."

"Watch his left hook."

"You've got this."

Young leaders desperately need power corners.

They need experienced men, who can help them recover after setbacks.

Who can encourage them after failures.

Who can remind them of truth when emotions threaten to overwhelm them.

The Apostle John wrote:

"I have no greater joy than to hear that my children walk in truth" (3 John 1:4, NKJV).

There is profound joy in watching those you've mentored, succeed.

Not because their success elevates you.

But because their success advances God's Kingdom.

Kingdom Success Unveiled

These four pictures reveal something profound.

You are not entering the final chapter of irrelevance.

You are entering the season of multiplication.

You are executing Kingdom succession.

An ambassador represents a kingdom.

He carries wisdom.

He builds bridges.

He influences outcomes.

He opens doors.

He develops leaders.

He prepares successors.

He advances purposes greater than himself.

This season is not about preserving your reputation.

It is about expanding your impact.

You are part of a much larger story.

Not your story.

His story.

God's story.

The masterpiece He has been painting since before you were born.

And your role now may be more significant than ever before.

So embrace this season.

Pass the baton.

Stabilize the ladder.

Offer backstage preparation and support.

Create a power corner.

The world needs what God has developed in you over decades.

The next generation needs your experience.

The Kingdom needs your investment.

You are not finished.

You are fruitful.

You are not being sidelined.

You are being repositioned.

You are not fading away.

You are advancing Kingdom agenda!

Your ceiling is becoming your successor's floor!

And perhaps, just perhaps . . .

The greatest season of your life has only just begun.

God is ready to use you ONE MORE TIME!

Chapter 14 Application

Four Pictures of Kingdom Succession

Reflect

Which of the four pictures best describes the season of life you are in today—**passing the baton, stabilizing the ladder, working backstage, or serving in the power corner?** Why does that picture resonate most with you?

Examine

Who are the people God has placed in your life, that need your wisdom, encouragement, and experience? How can you intentionally begin investing in them during this season?

Release

Is there a responsibility, title, position, or dream that God is asking you to place into someone else's hands? What fears make it difficult to let go, and how can you trust Him with the transition?

Encourage

Think about someone who is climbing the ladder behind you. What practical wisdom, relationship, opportunity, or encouragement could you offer, that would help them succeed with greater wisdom and fewer regrets?

Legacy

If your ceiling could truly become someone else's floor, who would you want that person to be? What intentional investment could you make today, that would help shape their future?

Pray

Ask God to give you the heart of a spiritual father. Pray for the humility to celebrate the success of others, the wisdom to know where to invest, and the courage to faithfully pass on what He has entrusted to you.

Act

Identify one younger leader, family member, friend, or colleague who you will intentionally invest in during the next month. Schedule a conversation, share your story, offer encouragement, or begin a mentoring relationship.

Final Reflection

Kingdom success is not measured by how high you climb, but by how many people you help climb after you. When your life is over, what do you hope others will say they were able to become because you invested in them?

Closing Challenge

Imagine standing before the Lord one day. Which would bring Him greater joy—that you built something impressive, or that you faithfully developed people who continued His work, long after you were gone? How will you begin living that legacy today?

CHAPTER 15

Don't Do It Alone

One of the greatest lessons I've learned is this: If you're going to experience God's strength one more time, you cannot do it alone.

When I think about Samson, I'm grateful that God gave him strength one more time. But I also can't help wondering how much pain could have been avoided if Samson had not lived such an isolated life. Everything revolved around Samson. There was no Nathan, no Jonathan, no Barnabas, no trusted people in his corner, who could speak truth into his life before disaster struck.

God never intended us to walk alone.

For me, that journey begins with Sandra.

Sandra is not just my wife.

She is my covenant partner. We've walked together. We've fought together. We've prayed together. We've cried together. We've believed God together. She's what people today might call a "ride or die" partner. She's the kind of person who isn't leaving when life gets hard. She's staying. She'll tell me the truth; she'll stand with me; and she'll fight for me, when I don't even have the strength to fight for myself.

Everybody needs people like that.

You need a power corner.

When we first moved to Atlanta, it was because I knew God had called us here. It wasn't our idea. It was His. After closing my martial arts school and leaving behind what had been such a big part of my life, we came to Atlanta because we believed God had told us to raise up a ministry.

The problem was . . . nothing was happening.

I couldn't get anything moving. Doors weren't opening. The ministry wasn't growing the way I thought it should. Every day seemed harder than the one before.

I remember walking into the house completely discouraged.

I dropped into a chair.

Sandra asked, "What's wrong?"

Like most men, I answered, "Nothing."

She knew better.

She walked over, grabbed me by my shirt collar, pulled me up out of that chair, and then pushed me right back down into it.

She pointed her finger at me and said, "God has a plan."

Then she started preaching to me.

She reminded me of everything God had spoken. She reminded me why we came. She reminded me that God had not changed His mind. Then she looked at me and said, "We're going to do this together. You're my man. You're going to accomplish everything God has called you to accomplish, and I'm going to be with you every step of the way."

I just started weeping.

Up until then, I really believed it all depended on me.

I thought, "I'm the man. I can make this happen. Faith works."

Faith does work.

But God was teaching me something I desperately needed to learn.

Faith works, but you were never meant to carry the assignment by yourself.

Years earlier, the Lord had already been trying to teach me that lesson. He had spoken to my heart that one could put a thousand to flight, but two could put ten thousand to flight. He had shown me that one plus one doesn't equal two in the Kingdom—it equals ten. He had told me that Sandra wasn't just my wife; she was supposed to be part of everything He had called me to do.

I believed it in theory.

I wasn't living it in practice.

That day, sitting in that chair, I finally understood.

I got up and thought, "I can make it another day."

Years later, after more than two decades of ministry, I found myself in another battle.

This one was even harder.

By then, I had accumulated plenty of battle scars. God had done wonderful things, but ministry had also wounded me.

A family left our church.

Today, I understand people come into our lives for different seasons. Some stay for a lifetime. Others stay only long enough to receive what God has for them, before He leads them somewhere else.

I didn't understand that back then.

When people left, I took it personally.

I thought they were my people.

They weren't.

They belonged to God.

Every time someone left, it felt like another knife in my heart.

I compared myself with other pastors. I questioned whether I was good enough. I wanted people to grow so badly, that when they stopped growing, I felt like I had failed.

Finally, I hit rock bottom.

I remember sitting on the side of my bed and saying to Sandra, "Maybe God never called me. Maybe I imagined all of this."

She looked at me with that same determination I'd seen years before.

She said, "Now wait just a minute."

"I was there."

"I know God called you."

"I know what He said."

"I know why we came here."

"And all you've ever wanted is to help people grow. That's all you've ever wanted."

Then she said something I'll never forget.

"Don't let people make you question what God has already settled."

She reminded me that Jesus told His disciples that when people would not receive their message, they were to shake the dust off their feet and move on.

Deliver the message.

Love the people.

But don't carry what God never asked you to carry.

That was my problem.

I wanted people to grow, more than they wanted to grow.

I was trying to hold on to the people who God was releasing.

Sandra reminded me of my calling, when I had forgotten it.

She believed for me, until I could believe again.

Little by little, my hope came back.

My vision came back.

My faith came back.

And I got up one more time.

If you're reading this chapter, let me tell you something.

When you decide to get up one more time, don't expect the enemy to applaud.

He's going to whisper,

"It didn't work before."

"You're going to fail again."

"What's the use?"

Those lies are designed to push you right back into the chair you just got out of.

Don't listen.

Listen to God.

Obey what He's saying today—not what fear is shouting.

Years ago, I knew a pastor who became so overwhelmed, that he believed death was his only escape. My heart still grieves when I think about him. The tragedy wasn't only that he died. The tragedy was that he believed the lie that there was no other way. Unfortunately, he got up one Sunday morning fully dressed and sent his family to church ahead of him, but he never showed up to preach. When his family returned home, they found him in his car in the garage. He took his life with a gun. He believed the lie that there was no other way.

There is another way.

God never intended you to carry the weight alone.

That's why you need a power corner.

You need people who know your calling.

People who know your heart.

People who know your weaknesses.

People who love you enough to tell you the truth.

People who will remind you what God has spoken, when you've forgotten.

But here's another lesson I learned.

Those relationships don't happen overnight.

Had I not spent years loving Sandra, encouraging Sandra, investing in Sandra, and walking with Sandra, she would not have been prepared to strengthen me when I needed her most.

The people who strengthen you tomorrow are usually the people you've invested in today.

So build those relationships now.

Encourage people.

Love people.

Walk with people.

Pray with people.

Because one day, you may need to borrow their faith until yours returns.

One more time isn't a solo journey.

You were never meant to finish alone.

Rebuilding My Capacity

In December of 2025, I hit a wall.

Everything on the outside looked successful. The ministry was growing. New opportunities were opening. God was blessing the work. But inside, I knew something wasn't right.

I had reached the limits of my capacity.

The vision God had placed in my heart was larger than the man I had become. If I was going to fulfill that vision, I couldn't simply work harder. I had to become someone with greater capacity.

That realization changed everything.

I recognized there were two things I needed to do.

First, I needed to work on myself.

Second, I needed to strengthen my power corner.

I'd heard it said many times, "If you're the smartest person in the room, it's time to get some more friends."

I already had friends. I had acquaintances. I had people I respected.

But I hadn't intentionally surrounded myself with the right people for the next season of my life. I needed men who could sharpen me, challenge me, and help me carry a vision that was larger than I could accomplish alone.

But before I could upgrade my relationships, I needed to upgrade myself.

So I asked a simple question:

What got me here?

What disciplines had God used throughout my life, that I needed to reclaim and strengthen?

As I reflected, three words emerged, that described who I am at my best.

I call them **PMA:**

Psalmist. Martial Artist. Author.

These aren't simply activities I do.

They describe the man God created me to be.

Psalmist

The first part of PMA is the Psalmist.

For me, a Psalmist is someone who lives immersed in the presence of God.

It's worshiping.

Praying.

Writing songs.

Singing new songs to the Lord.

Playing instruments.

Listening.

Waiting.

Living with a sensitivity to the Holy Spirit and His supernatural work.

Years ago I played an Omnichord, but I eventually stopped because I wanted to learn piano and guitar. I realized it wasn't really about learning instruments—it was about cultivating worship.

Jesus told His disciples not to rush into ministry. He told them to wait in Jerusalem until they were *"clothed with power from on high"* (Luke 24:49, NIV).

Power always comes before purpose.

Like a tree planted by rivers of water, fruit doesn't come from striving. Fruit grows out of abiding.

That's the heart of the Psalmist.

Martial Artist

The second part of PMA is the Martial Artist.

Closing my Martial Arts school didn't mean I stopped being a martial artist.

Somewhere along the way, I lost that disciplined version of myself.

I gained weight.

I became less flexible.

There was a time I could do full splits, jump kicks, and demanding forms. Those days may never fully return, but the discipline that produced them can.

So I started rebuilding.

I hired a wellness coach.

I committed to losing weight.

I returned to my forms and daily training—not to prepare for tournaments, but to prepare for purpose.

One of my daughters has asked me to teach her martial arts, and I plan to teach my grandson as well.

I'm rediscovering the mindset that shaped me for decades: discipline, focus, consistency, and continual growth.

The Martial Artist isn't just something I used to be.

It's still part of who I am.

Author

The third part of PMA is the Author.

Writing has always been one of God's gifts in my life.

But gifts don't grow without discipline.

I realized I needed a regular rhythm of creating—thinking, writing, teaching, and building.

That's why this book matters so much to me.

It's not simply another project.

It's part of my own transformation.

Every chapter I write is helping me become the man who God is calling me to be.

Becoming Before Building

PMA is my capacity building plan.

Not because I wanted another system.

But because I understood something I had forgotten.

God expands our influence by first expanding our capacity.

Before He increases what flows through us, He

increases what He can accomplish in us.

Only then was I ready to strengthen the second part of my preparation . . .

My power corner.

You need a power corner.

One of the greatest illustrations of this happened in boxing.

When Mike Tyson fought Buster Douglas, Tyson wasn't just fighting the man across the ring. He was fighting everything that had been going on in his own life. By that point, he had separated himself from the people who had helped shape him into a champion. The trusted voices that had challenged him, corrected him, and prepared him were gone.

Even his corner showed it.

When his eye began to swell, his team wasn't prepared. Instead of having the proper equipment like an enswel to bring the swelling down, they improvised with a rubber glove filled with ice water. It was obvious they weren't the experienced team that had once surrounded him.

Here's my point.

Champions don't just need talent.

Champions need the right corner.

Between every round, a fighter sits down. He's tired. He's been hit. Sometimes he can't even see clearly because of the swelling. That's when his corner goes to work. They wipe away the blood. They help him see what he can't see. They remind him of the game plan. They tell him where he's winning and where he's vulnerable. Then they send him back into the fight.

Life is no different.

You're going to have rounds where you've been hit hard. There will be days when you're exhausted, discouraged, and wondering if you can keep going for one more round.

That's when you need a power corner.

You need people who know you well enough to tell you the truth.

People who can see what you can't see.

People who remind you of what God has said, when all you can hear are the lies of the enemy.

Don't ever make the mistake of thinking you can finish this race by yourself.

Find your power corner.

Build your power corner.

Protect your power corner.

Because one more time is almost always the result of somebody standing in your corner saying, "Get up. You're not finished yet."

Now let me give you another image of the right power corner.

Think about the great fight between Sugar Ray Leonard and Thomas Hearns.

Sugar Ray was taking a tremendous beating. His eye had been badly damaged. From the outside, it looked like the fight was slipping away.

But Sugar Ray had something working for him that many people overlook.

He had the right power corner.

Standing in that corner was the legendary trainer Angelo Dundee, the same man who helped train Muhammad Ali. Angelo wasn't just treating Sugar Ray's eye. He was treating Sugar Ray's mind.

As Sugar Ray sat on that stool between rounds, discouraged and exhausted, Angelo leaned over and challenged him.

"You're blowing it! You're blowing it now, son! You're blowing it!"

He wasn't trying to tear him down.

He was trying to wake him up.

He was reminding him of who he was.

He was putting courage back into him.

He was restoring his vision.

He was rebuilding his confidence.

He dealt with the swelling around Sugar Ray's eye, but he also dealt with the heat in the ring, the weariness in his arms, and the discouragement that had settled into his heart. According to Sugar Ray, those words from Angelo were like electricity to his body. He got up and knocked Tommy Hearns down and out of the ring. It was the first Tommy Hearns was knocked down in his career! Sugar said that Angelo's words were the perfect thing he needed to hear.

That's what a real power corner does.

They don't just help you with the problems everybody can see.

They help you with the battles nobody else knows you're fighting.

They remind you of who God called you to be.

They remind you of what God has spoken over your life.

They remind you that one bad round doesn't determine the outcome of the fight.

Then they look you in the eye and say,

"Now get back out there."

"You're not finished yet."

Every one of us needs people like that.

When I think back over my own life, Sandra has been that person for me more than anyone else. There were moments when I was ready to quit, when I questioned my calling, when I wondered if I had anything left to give. God used her to put courage back into me, to remind me of His promises, and to send me back into the fight.

That's a power corner.

So let me ask you a question.

Who's in your power corner?

Build Your Power Corner

At this stage of my life, I've intentionally stacked my corner.

Of course, Sandra is the most important person in that corner, but she's not the only one. Over the years, I've become very strategic about surrounding myself with the right people.

There came a point when I realized I had become the person everyone else depended on, but I didn't have enough people who I depended on.

The answer wasn't that I needed to work harder. It was that I needed more wisdom than I possessed. I needed people who had already walked the road I was now traveling.

Not just successful people.

Not just older people.

People with specific experience in the exact areas where I had weaknesses.

So I began intentionally adding people to my corner.

One person helps me think strategically.

Another helps me build organizational systems.

Another challenges my leadership.

Another speaks into my spiritual life.

At the time of this writing, I'm still doing this. Every quarter, I sit down with men who have greater experience than I do in specific areas. I tell them where I am. I let them ask difficult questions. Then I listen.

Really listen.

That's the hard part.

It's one thing to invite wise people into your life. It's another thing to receive their correction without becoming defensive.

A power corner only works if you trust the people you've invited into it.

You must allow them to speak honestly.

You must give them permission to challenge your assumptions.

You must let them point out blind spots you cannot see.

The truth is, every one of us has deficiencies. We all have weaknesses. We all have areas where someone else is stronger than we are.

Great leaders don't pretend those weaknesses don't exist.

They build teams that compensate for them.

The goal isn't to become good at everything.

The goal is to become wise enough to know where you need help.

Your power corner begins at home. For me, it begins with Sandra. She knows me better than anyone else, and her wisdom has protected me more times than I can count.

But even Sandra would tell you she isn't called to help me in every area.

God never intended one person to meet every need.

That's why He places other gifted people around us.

As you prepare for your "one more time," ask yourself:

Who is in your corner?

Who has permission to challenge you?

Who has experience you don't have?

Who tells you the truth instead of simply telling you what you want to hear?

And perhaps the most important question of all:

When they speak . . . do you listen?

Your next season may not depend on working harder.

It may depend on listening better.

Chapter 15 Application

Don't Do It Alone

Reflect

Who is currently in your "power corner"—the people who know you well enough to encourage you, challenge you, and remind you of God's calling when you begin to lose heart? How can you strengthen those relationships?

Remember

Think about a time when someone spoke life into you during a season of discouragement. What did they say or do that helped you get up "one more time"? How did God use them in your life?

Examine

When you're struggling, are you willing to let others see your weakness, or do you tend to carry the burden alone? What keeps you from asking for help?

Invest

Who are you intentionally encouraging and strengthening today? The people you invest in now may become the very people God uses to strengthen you in the future. How can you pour into them this week?

BUILD

If you were intentionally building your power corner, who would you invite into it? What kinds of people do you need—someone who challenges you spiritually, someone who offers wise counsel, someone with experience you lack, or someone who simply believes in you?

LISTEN

When trusted people speak correction or encouragement into your life, how do you typically respond? Are you teachable, or do you become defensive? What would it look like to receive wise counsel with humility?

PRAY

Ask God to surround you with people who will speak truth in love, encourage your faith, challenge your blind spots, and strengthen you for His calling. Pray also that He would make you that kind of friend, mentor, and encourager for others.

ACT

Reach out to one person in your power corner this week. Thank them for their investment in your life, ask for their honest perspective on an area where you want to grow, and intentionally listen to their counsel.

FINAL REFLECTION

No one finishes God's calling alone. Who has God placed in your corner to help you finish well? And just as importantly, whose corner has God called you to stand in?

Closing Challenge

One more time is rarely the result of individual determination alone. More often, it happens because someone looked you in the eye and reminded you of who God called you to be. Whose courage will you restore this week?

Encourage

Who in your life may be sitting in discouragement today? Whose faith is weak? Whose vision has become clouded? How could God use your words, your presence, or your prayers to remind them of His promises, and help them get up one more time?

Grow

I remind you that your next season often requires greater capacity. What area of your life is God inviting you to strengthen—your relationship with Him, your character, your health, your leadership, your knowledge, or your relationships? What one intentional step can you take this week to grow?

Closing Prayer

Father, thank You for never intending me to walk alone. Thank You for the people You have placed in my life to encourage, challenge, and strengthen me. Give me the humility to receive their wisdom and the courage to be that kind of friend, mentor, and encourager to others. Expand my capacity, deepen my faith, and surround me with people who will help me finish well. And when someone around me grows weary, use my words to restore their courage and remind them that You are not finished with them yet. In Jesus's name, Amen.

CHAPTER 16

A Conversation With My Wife, Sandra

One of the greatest mistakes a man can make is assuming that God's call on his life belongs only to him.

It doesn't.

God designed marriage as a partnership.

Before you launch into your "one more time" assignment, bring your wife into the journey.

I asked my wife, Sandra, to share her perspective after nearly five decades of marriage. And she shared:

"We've been married for almost forty-seven years.

From the very beginning of our relationship, I

sensed that John was called into ministry.

I didn't fully understand what that meant or what my role would be, but God had already placed something in my heart.

He prepared me to support my husband.

Supporting John has never been about simply helping a man.

It has been about saying 'yes' to God.

I believe the call of God on a husband is also a call upon the household.

It includes the wife.

It includes the children.

In order for a husband to fulfill God's assignment, he needs the support of his family.

We need to be in agreement, not in opposition.

Scripture says the two shall become one.

If we're truly one, then we're moving in the same direction ... believing together . . . serving together . . . and building together.

Imagine a husband giving everything he has to God's call, while his wife is moving in a completely different direction.

The family is divided.

And division weakens the mission.

Unity strengthens it.

I wanted my life to honor God.

That meant embracing my role as John's wife and supporting the vision God had placed in his heart."

Sandra then turned directly to wives.

"Ladies . . . your husband needs your support.

I know it's easy to become consumed with children . . . grandchildren . . . activities . . . and all the responsibilities of everyday life.

Those things matter.

But they must never replace Christ as the center of your life.

Sometimes we unknowingly make our family our highest priority instead of making Jesus our highest priority.

When Christ is first, He helps us walk together as one.

The goal isn't simply agreeing with your husband because he's your husband.

The goal is seeking God's will together."

I then asked Sandra a question many women have wondered.

"What about the wife who says, 'What about my dreams? I've spent thirty years supporting my husband. Doesn't God have something for me too?'"

She smiled before answering.

"Absolutely He does.

I spent years helping John discover and build the ministry God entrusted to him.

Ministry wasn't built overnight.

There were seasons of learning . . . growing . . . making mistakes . . . and trusting God through every step.

I was there through it all.

Just as I supported John's calling, he faithfully encouraged me to pursue the calling God had placed on my life.

There were seasons when I wanted to quit ... when discouragement made me question whether I should continue.

But John never let me give up.

He would lovingly encourage me to get back in the fight.

He even wrote a little song that he would sing whenever I felt discouraged about what God was calling me to do.

It always reminded me that God wasn't finished with me.

Then, as we entered this new season John has been describing, something beautiful happened.

As his ministry became firmly established, God began speaking to my heart about ministering to women through Christian Business Women's Connection (CBWC).

What I discovered is that God's calling on a marriage is often seasonal.

Sometimes one spouse carries the greater weight while the other provides strength, encouragement, and support.

Then the seasons change.

The one who was leading steps alongside to encourage the other.

That's the beauty of a covenant marriage.

We are not competing with one another—we are

completing one another.

We take turns strengthening each other so that together we can fulfill everything God has called us to do.

Because I had spent years helping build CBMC alongside John, he now had the wisdom, experience, and systems to help me build CBWC.

I didn't have to start from scratch.

I was able to stand on everything we had learned together.

God will never ask you to faithfully support your husband without also caring about the calling He has placed inside of you.

Your calling may look different.

It may not even be a women's ministry.

But as you build together, you'll discover that your husband becomes one of your greatest supporters."

At that point I wanted to add something from a husband's perspective, and so I said:

"Men . . .

This doesn't happen automatically.

One day I asked Sandra when she was seventy-three,

'When you're eighty years old, what do you want your life to look like?'

I wanted to know what God had placed in her heart during this season.

Just as she spent decades helping me fulfill God's assignment . . .

I knew it was now my responsibility in this season to help her fulfill hers.

In fact, I sensed the Lord clearly saying to me,

'Help Sandra.'

That's exactly what we're doing today.

As the message of *One More Time* begins to stir your heart . . .

Don't run ahead by yourself.

Build this vision together.

It's much easier for a wife to support her husband's calling when she knows she is deeply loved . . .

deeply valued . . .

and intentionally included.

Leadership isn't demanding support.

It's earning trust.

Throughout our marriage, I've always protected time with Sandra.

Early on it was a monthly date.

Then it became twice a month.

Eventually it became a weekly date night.

Today, for many years now, we've enjoyed a weekly Date Day.

That rhythm tells her,

'You matter more than my ministry.'

If ministry causes you to neglect your marriage . . .

You've misunderstood ministry.

Men . . .

Schedule the date.

Protect the date.

If work interrupts it occasionally . . .

Make it up immediately.

Your wife should never have to compete with the calling God gave you.

Bring her with you.

My final thoughts . . .

I'm going to say it . . .

One more time.

Men . . .

Don't run off without your wife.

I know what happens.

You're inspired.

God has stirred something deep inside of you.

You're ready to charge into hell with a water pistol.

You're saying,

'God, strengthen me one more time!'

You feel like old Caleb declaring,

'Give me this mountain!'

And that's a beautiful thing.

But don't let your excitement cause you to leave your wife in the dust.

Too many men spend years working long hours . . .

building careers . . .

and chasing success . . .

while their wives faithfully stand beside them.

Then, when God awakens a new calling . . .

the husband expects his wife to instantly drop everything and catch up to a vision she hasn't had time to process.

Then, if she struggles . . .

or hesitates . . .

he blames her.

'She just didn't want to obey God.'

No, brother.

Slow yourself down.

God didn't call you to fulfill your purpose at the expense of your marriage.

He called you to walk in purpose with your wife.

Take time to bring her into the journey.

Talk about what God is doing in your heart.

Pray together.

Take a weekend away.

Go on a retreat.

Dream together.

Listen to her fears . . .

as much as you share your faith.

Make sure she feels valued . . .

before she feels enlisted.

As I'm writing this, Sandra and I just had our regular Date Day yesterday.

This morning I fired up the grill . . .

cooked hot dogs with all the toppings . . .

and we sat in our garage eating hot dogs for breakfast.

It wasn't expensive.

It wasn't glamorous.

It probably wasn't even the healthiest breakfast we've ever eaten.

We had two fans blowing because it was so hot in the garage.

While we were eating and enjoying each other . . .

I looked at Sandra with a cheesy smile and said,

"Sandra . . . this is high living!"

We both laughed.

But it wasn't about the hot dogs.

It wasn't about the garage.

It wasn't even about breakfast.

It was about being together.

Those simple moments communicate something every wife longs to hear.

You still matter.

You're still my girl!

You're still my priority.

If I had it to do all over again, I would still choose you.

The greatest ministry you will ever have . . .

is not the one that applauds you from a platform.

It's the covenant you nurture at home.

If God has given you one more assignment . . .

make sure you invite your wife into the journey.

Walk together.

Pray together.

Dream together.

Don't just finish your race.

Finish it hand in hand with the woman God entrusted to walk beside you.

Don't do it alone.

Do it together. And enjoy the commanded blessing that comes with unity.

Chapter 16 Application
A Conversation With Wife, Sandra

God never intended your "one more time" assignment to be a solo journey. Marriage is one of His greatest gifts, and when a husband and wife walk together in unity, they experience God's favor in a unique way. Take time this week to intentionally strengthen your partnership and seek God's direction together.

For Husbands
Have I truly invited my wife into the calling God has placed on my life, or have I simply expected her to follow? Does my wife know she is more important to me than my work or ministry? Have I asked her what dreams and desires God has placed in her heart? Am I protecting regular, intentional time with her? If someone observed our marriage, would they see unity, love, mutual honor, and partnership?

For Wives
Am I praying faithfully for my husband's calling and encouraging him to become all God created him to be? Have I shared my own dreams and desires with my husband in a loving and healthy way? Are there fears, disappointments, or unmet expectations that we need to discuss openly? Am I seeking Christ first so that our marriage remains centered on Him above everything else?

For Both of You

Are we pursuing God's purpose together? What dream has God placed in our hearts that we can pursue as one? What one step can we take this month to strengthen our marriage? Is there any unresolved hurt, misunderstanding, or disappointment that needs to be healed through prayer, forgiveness, and honest conversation?

This Week's Challenge

Schedule an uninterrupted date this week. Turn off your phones. Go for a walk, sit over coffee, share a meal, or simply spend time talking together.

Ask each other these three questions:

1. What is God doing in your heart right now?
2. How can I better support you?
3. What dream can we pursue together over the next year?

Then take a few moments to pray together before you leave.

Don't worry about solving every problem. The goal isn't perfection—it's connection. Great marriages aren't built through one grand moment. They're built through thousands of intentional conversations, shared prayers, and faithful acts of love.

A Prayer

Father, thank You for the gift of marriage. Thank You for bringing us together as one. Help us to walk in unity as we pursue the calling You have placed on our lives. Give us hearts that listen before speaking, encourage before criticizing, and serve before expecting to be served. Teach us to honor one another, to protect our covenant, and to keep Christ at the center of our home. As You give us "one more time" to serve Your Kingdom, may we never run ahead alone, but walk hand in hand, trusting You every step of the way. In Jesus's name, Amen.

One More Time Principle

Your greatest legacy will not simply be the ministry you build, the business you lead, or the lives you influence. It will be the covenant you faithfully nurture at home.

If God gives you one more assignment, don't answer His call alone. Answer it together.

"Behold, how good and pleasant it is when brothers dwell in unity! . . . For there the LORD has commanded the blessing, life forevermore." — (Psalm 133:1, 3, ESV).

"I appeal to you, brothers, by the name of our Lord Jesus Christ, that all of you agree, and that there be no divisions among you, but that you be united in the same mind and the same judgment" (1 Corinthians 1:10, ESV).

CHAPTER 17

Final Commission

Before you close this book, I want to leave you with one final challenge.

Your story is not over.

Your assignment is not finished.

God has not preserved you simply so you could grow older.

He has preserved you so you can grow wiser, stronger, and more influential.

Every scar you've earned . . .

Every battle you've survived . . .

Every lesson you've learned . . .

Every tear you've cried . . .

Every victory you've celebrated . . .

God intends to use them all.

There is a younger generation waiting for someone to believe in them.

Someone to encourage them.

Someone to teach them.

Someone to walk beside them.

Someone to show them what it looks like to finish well.

Be that man.

Build yourself.

Build your power corner.

Transfer your wisdom.

Raise up successors.

Leave a legacy that will outlive you.

And every morning, when your feet hit the floor, let this be your prayer:

Lord, strengthen me one more time.

One more conversation.

One more disciple.

One more act of courage.

One more risk of faith.

One more leader developed.

One more life changed.

One more generation impacted.

One more time.

And when your race is finally finished, may you be able to say, like Paul:

"I have fought the good fight, I have finished the race, I have kept the faith" (2 Timothy 4:7, NIV).

May your greatest days not simply be behind you.

May your greatest impact still lie ahead.

Rise up.

Answer the call.

Finish well. One More Time.

Chapter 17 Application
Final Commission

Reflect
Looking back over your life, what scar, victory, lesson, or season of growth has God uniquely prepared to use in this next chapter of your life?

Commit
Who is one person God is calling you to encourage, mentor, disciple, or simply walk alongside? What is one intentional step you will take to begin investing in them?

Pray
Will you make this your daily prayer?

"Lord, strengthen me one more time."

Ask Him each day for the strength to love one more time, forgive one more time, serve one more time, lead one more time, and invest in one more life for His glory.

Act
What is your "one more time" step? Is it a difficult conversation, an act of forgiveness, a step of faith, a renewed commitment, or an opportunity to serve? Whatever God is asking of you, choose one act of obedience and take it this week.

Finish Well

If your greatest Kingdom impact still lies ahead, what priorities, habits, or relationships must you cultivate today, so that you can finish your race with faithfulness, courage, and joy?

Final Reflection

Your story is not over. Your assignment is not finished. God has strengthened you before, and He is able to strengthen you again.

How will you live your "one more time"?

CHAPTER 18

FINAL PRAYER

My Growth Partner,

Thank you for taking the time—your valuable time—to read this book and to allow me the privilege of sharing my story with you. More than anything, I hope I've shared my heart.

I am deeply passionate about reaching the next generation. If this book can inspire men like you to step into this Kairos moment and invest your life in others, then God has done far more than I ever imagined.

When the Lord first gave me this assignment, I thought it was simply something He wanted me to do. Little did I know that He intended to multiply it through the lives of other men. Today, I have the privilege of standing alongside faithful men here in

Atlanta and across the country, who have embraced this same calling.

Now, my prayer is that God will multiply it through you.

Would you allow me the privilege of praying for you?

Father God,

I come to You in the mighty name of the Lord Jesus Christ.

First, I thank You for my Growth Partner who has read these pages. Thank You for the courage You have placed within him to answer Your call and to become part of Your greater story.

Lord, You have been preparing him for this moment all along.

You have walked with him through trials, disappointments, victories, failures, heartbreaks, and triumphs. You have carried him through seasons when he didn't think he could take another step. Yet through it all, You have remained faithful.

Your Word declares that through faith and patience, we inherit the promises.

Thank You that he has not walked away.

There were moments when he wanted to.

There were seasons when he was tired.

Times when he questioned.

Times when he doubted.

Yet here he stands.

Like Samson . . .

Like Peter . . .

Like Paul . . .

Still standing.

Still believing.

Still willing to say, "Lord, use me."

Father, together we now report for duty as soldiers in Your Kingdom.

Your Word tells us to endure hardship as good soldiers of Jesus Christ and not become entangled in the affairs of this life, so that we may please the One who enlisted us.

Lord, that is our desire.

We want to please You.

We surrender ourselves afresh before You today.

I pray that as my brother submits himself to Your grace, Your Spirit, and Your purposes, that Your presence would rest upon him with fresh power.

Increase his capacity.

Strengthen his inner man.

Heal every wound.

Redeem every regret.

Use every scar.

Turn every disappointment into wisdom.

Turn every failure into testimony.

Turn every victory into a platform for serving others.

Lord, surround him with the right people.

Build his power corner.

Bring faithful men into his life who will sharpen him, encourage him, challenge him, and hold him accountable.

Help him not only to finish—but to finish well.

May this be the most fruitful season of his life.

May he accomplish more than he ever dreamed possible.

May he invest in the next generation.

May he build successors.

May he leave a legacy that points people to Jesus Christ, long after he has gone home to be with You.

And Father, just as Samson prayed . . .

I pray over my friend today:

Strengthen him one more time.

Strengthen him to love one more time.

To believe one more time.

To lead one more time.

To forgive one more time.

To mentor one more time.

To build one more time.

To preach one more time.

To serve one more time.

To obey one more time.

And through that one more time, may he bring You more glory than at any other season of his life.

We ask all of this in the mighty, powerful, and matchless name of Jesus Christ.

Amen.

Chapter 18 Application

Final Prayer

Receive

As you prayed this final prayer, what words or phrases seemed to speak directly to your heart? Why do you believe God highlighted them for you in this season of your life?

Surrender

What area of your life needs God's strength today? Where are you asking Him to "strengthen you one more time," so you can continue His calling with renewed faith and courage?

Believe

Looking back over your life, where have you seen God turn disappointment into wisdom, failure into testimony, or pain into purpose? How does His faithfulness in the past give you confidence for the future?

Invest

Who has God placed in your life that needs your encouragement, wisdom, or friendship? What is one intentional step you will take this week to invest in them?

Commit

As you close this book, what commitment are you making before God? Write it down. Pray over it. Then begin living it out one day at a time.

Final Prayer

Before you turn the page, pause for a moment. Quiet your heart before the Lord and make this your own prayer:

Father, strengthen me one more time. Use every lesson, every scar, every victory, and every failure for Your glory. Help me to finish well, invest in the next generation, and faithfully complete the assignment You have given me. Whatever years You have entrusted to me, may they become my most fruitful years. In Jesus's name, Amen.

Final Reflection

Your story is still unfolding.

Your calling is still alive.

Your greatest investment may still be ahead of you.

So tomorrow morning, when your feet touch the floor, begin the day with one simple prayer:

Lord . . . strengthen me one more time.

CHAPTER 19

A Final Benediction

My friend, if you have prayed that prayer with sincerity, I believe heaven has heard you.

I believe God is not finished with you.

I believe your scars have prepared you.

I believe your experiences have qualified you.

I believe your wisdom is desperately needed.

There is a younger generation waiting for someone just like you.

They don't need perfection.

They need authenticity.

They don't need a superhero.

They need a spiritual father.

They don't need someone with all the answers.

They need someone willing to walk beside them.

So don't spend the rest of your life protecting what you've built.

Spend it investing in who God has called you to build.

One day you'll stand before the Lord, and it won't matter how much money you accumulated, how many titles you held, or how many accomplishments filled your résumé.

What will matter is this:

Did you know Him?

Did you obey Him?

Did you finish your race?

And did you leave behind men and women who loved Jesus more, because you answered His call?

That's my prayer for you.

So stand up.

Take your place.

Answer His call.

Build yourself.

Build your power corner.

Invest in the next generation.

Leave a legacy that eternity will celebrate.

And when the day comes that you see Jesus face to face, may you hear the words every faithful servant longs to hear:

"Well done, good and faithful servant" (Matthew 25:21, ESV).

"The LORD bless you and keep you; the LORD make his face to shine upon you and be gracious to you; the LORD lift up his countenance upon you and give you peace" (Numbers 6:24–26, ESV).

May Adonai use you.

May Abba multiply your influence.

May Yahweh grant you great victories in your amazing future.

And through El Shaddai, may your life forever declare:

ONE MORE TIME

Chapter 19 Application

A Final Benediction

Reflect

Looking back over your life, which scar, experience, victory, or lesson do you now recognize as God's preparation for this season? How might He use your story to encourage someone else?

Invest

Who is one person God is calling you to encourage, mentor, disciple, or simply walk beside? What first step will you take to begin investing in their life?

Evaluate

When you one day stand before the Lord, what do you hope will matter most? Are there ambitions, distractions, or accomplishments you need to surrender so you can focus on what has eternal value?

Respond

What one act of faith, obedience, or love is God asking you to take today? Don't wait for the perfect moment. Take the next faithful step.

Live Your Legacy

If those who know you best could describe the legacy of your life in one sentence, what would you want that sentence to be? What choices can you make today that will help make that legacy a reality?

Final Reflection

One day, every title will fade.

Every accomplishment will be forgotten.

Every earthly success will pass away.

But the lives you touched . . .

The people you loved . . .

The leaders you developed . . .

The souls you pointed to Christ . . .

Those will remain for eternity.

When your race is finished, what do you hope will still be living because you answered God's call?

BIOGRAPHY

John brings a distinct mix of ministry, business, and leadership expertise to his work. His career spans over 45 years of developing human capital across diverse social, ethnic, technical, and educational environments. Currently, he serves as the Metro Atlanta Area Director for the Christian Business Men's Connection (CBMC)—a ministry founded in 1930, that now impacts 98 countries and 330 US cities.

John is a certified trainer, coach, speaker, DISC consultant, and founding member of the John Maxwell Leadership Certified Team. He holds an associate degree in electrical engineering, as well as a bachelor's, master's, and doctorate degree in pastoral theology. Committed to lifelong learning, he has earned numerous certificates and certifications ranging from business management and leadership to life coaching and streaming media consulting. Guided by the philosophy that the key to fulfilling your dreams is to grow into them, John also holds a fifth-degree black belt (master level) in Moo Duk Kwan Tang Soo Do martial arts with the National Black Belt Karate Association.

John possesses a unique gift: the ability to impact "the one" who impacts the many. Whether working with a pastor, a marketplace minister (CEO), a community leader, a father, or a mother, he believes everyone has a God-given purpose to make a difference in this world. His life mission is to help you realize that difference.

Think of John as Morpheus to Neo, Angelo Dundee to Muhammad Ali, or Professor X to the X-Men. He builds capacity in passionately gifted people, empowering them to achieve the dreams God has given them. Rather than seeking the spotlight himself, John is the strategist behind the scenes—supporting, strengthening, and maximizing the potential of the leader on the stage.

John and his wife, Sandra (married since 1979), are the proud parents of three daughters, Rhea, Tasha, and Alexus, and the *very* proud grandparents of Zoe Alexandria, Caelix Jonathan, and Sophia Lee.

FULL ENDORSEMENTS AND PRAISE FOR *ONE MORE TIME*:

"Some men reach the back half of life and quietly decide the fight is over. They mistake gray hair for a discharge paper. I spent a career standing near men at the height of their power, and I watched others walk off the field with rounds still in the magazine. John Posey wrote *One More Time* for the ones who walked off, and he will not let them stay seated.

"He tears down the lie that your best days are spent and God has handed the ball to someone younger. John has the scars to back up every page. I once told a seasoned businessman in Singapore, too bruised to believe he had anything left to give, that his survival was the point. The young men around him were going under, and his scars were the life preserver. John says it plainer than I did: 'Do you have any scar tissue? Then you're qualified.'

"The first half of life builds success. The second half builds successors. John has lived every word of this through surrender, forgiveness, and a calling he had to rebuild from the ground up. He will help you live it too. Read it. Then go pull someone out of the water. One more time."

- Christopher C. Simpson, President & CEO, CBMC International, US Secret Service (Ret.), US Marine Corps

Christopher C. Simpson is President and CEO of CBMC International, the world's largest evangelical marketplace ministry, serving in more than ninety nations.

"John Posey writes out of the overflow of his life. He models how to finish well with his intentionality of investing his time and wisdom in faithful, available, and teachable young men. *One More Time* is an inspiration for all of us to pour into the next generation, and in the process, as Paul says to fight the good fight, to finish the course, and to remain faithful, until we go to be with Jesus."

- Boyd Bailey, President, NCF Georgia | Founder, Wisdom Hunters

"CBMC Metro Atlanta Area Director John Posey asked me to write a brief endorsement for his book, *One More Time*. I realized after looking at his manuscript that the best way I can affirm the value of this book is to use John's own description: '*One More Time* is a book about finishing well, discovering God's purpose in every season of life, and investing in the next generation. It's written especially for men who may be wondering if their greatest days are behind them. My hope is to encourage them to believe that God is not finished with them and that their greatest Kingdom impact may still lie ahead.'

"That rich summary is characteristic of the personal, practical, and pithy statements that permeate this little guidebook. John's reflections on Chronos and Kairos, and condemnation to righteousness are dynamic and practical, and they are driven home in the Reflection & Action exercises at the end of each chapter. I recommend this book to followers of Jesus who want to hear 'Well done' when they finish their race."

- Kenneth Boa, Founder, Reflections Ministries and Museum of Created Beauty

"I wholeheartedly endorse your book because it is much needed as a guide for those of us at this stage of our journey. More importantly, it helps those around us understand our challenges so they may encourage us to 'finish well.' Lastly, I know the book's truth, having been blessed to witness your ministry and mentorship. I pray that God will bless you as you share this work with many 'one more timers'!"

- Charles W. Buffington, CEO and Founder, Sonseeker Media Partners

"John Posey's book offers a compelling reminder that every life has purpose, every scar can carry meaning, and it is never too late to finish well. With biblical encouragement, practical wisdom, and a message of hope, it inspires readers to trust God's plan and embrace the next chapter of their journey with faith and confidence."

- Dr. Terence Chatmon, Victorious Family

"*One More Time* is a timely call to a Kairos moment for believers who have faithfully served the Lord and now find themselves entering the later chapters of life and ministry. John Posey reminds us that these years are not for retreat but for renewed purpose.

"I believe this is one of the most important messages for our generation. Many Christians today have more time, more resources, and more wisdom than they've ever had before, yet too often those gifts are spent on themselves. *One More Time* calls us to recognize God's invitation, join Him in His work, and pray, like Samson, 'Lord, strengthen me one more time' for what may be our most fruitful season yet.

"I've had the privilege of watching my friend John

Posey live this message. I thoroughly enjoyed reading this book, but I've been even more encouraged by seeing its truths embodied in his life. This isn't simply a message he teaches—it's a life he lives. I highly recommend *One More Time* to anyone who wants to finish well and discover that God may still have some of His greatest work ahead."

- Jason Dees, Senior Pastor, Christ Covenant Church

"John Posey's *One More Time* is more than the story about one man's life—it is a testimony to God's faithfulness and His ability to use ordinary people to make an extraordinary impact on others. Through biblical stories and personal experiences, John encourages readers to reflect on their own journey, learn from life's lessons, and move forward with purpose and conviction.

"In an era focused on transferring wealth, influence, and opportunity, he reminds us that our greatest responsibility may be the transfer of spiritual wisdom and faith. Through practical applications and thoughtful reflection questions, he equips readers not only to grow personally but also to invest intentionally in others.

"This book is a powerful illustration of Ephesians 2:10—that we are God's workmanship, created in Christ Jesus for good works that He prepared in advance for us to do. As John recounts God's hand at work throughout his life, readers will be inspired to discover and faithfully pursue God's purpose for their own.

"If you desire a book that will strengthen your

faith, challenge your perspective, and inspire you to leave a lasting legacy, *One More Time* is a book you'll want to read—and then share with someone else. John's story is proof that a life surrendered to God can leave an eternal impact on generations yet to come."

- Andre Kennebrew, Director of Relationships, Lifeshape

"I have watched John Posey do this book before he wrote it. In CBMC, I saw him hand batons to men half his age, hold ladders he could have been climbing, and refuse the applause he'd earned. *One More Time* is that life in print—honest about his own failures, clear-eyed about the moment, and generous with what he's learned. He names something I encounter in nearly every organization I work with: The wisdom in the room is about to walk out the door, and almost no one has a plan for transferring it. We are moving wealth without moving wisdom, and everything we've gained sits one generation from extinction. John's question, 'Could your ceiling become someone else's floor,' is worth building the rest of your life around. If you're older, scarred, and quietly wondering whether God is finished with you, his answer is bracing: If you're still here, He isn't."

- Lee Truax, President, DiscipleMaking Mission, and past President, CBMC

"By honestly admitting his own failings and the lessons they taught, John masterfully illustrates the power and imperative of a life that is fully God-led, God-dependent, and God-honoring. We were created in the image of Christ. This book beautifully illustrates the transformation journey to fully becoming His likeness."

- Robert W. (Bob) Lewis, President, Lewis Leadership LLC

"Leading the national launch of CBMC, SageMen has convinced me that there is an untapped army of seasoned believers whose greatest days of Kingdom impact may still be ahead. John Posey's *One More Time* captures that vision beautifully. I've known John as a man whose life reflects the message he writes. He reminds us that God never wastes a surrendered life—our scars become our ministry, our weakness becomes our witness, and our story becomes hope for someone else. Every believer who wants to finish well and leave a legacy that outlives them will find this book both inspiring and deeply encouraging."

- Alan Smith, Regional Director - West Region Christian Business Men's Connection

"Every man is a runner by nature. We are either running 'from something' or 'to something.' With *One More Time*, John reminds and encourages men to keep running to what matters in the long run. Our design and calling are to love God and love others. One of the greatest opportunities for this lies in the generations behind us. God has called John into His purpose, and he's inviting us to join him in doing the same and finish strong. Press on! *'No, dear brothers and sisters, I have not achieved it, but I focus on this one thing: Forgetting the past and looking forward to what lies ahead, I press on to reach the end of the race and receive the heavenly prize for which God, through Christ*

Jesus, is calling us' (Philippians 3:13-14, NLT)."

- Sean Robertson, Director of Ministry, CBMC

www.ingramcontent.com/pod-product-compliance
Lightning Source LLC
LaVergne TN
LVHW051926051126
851857LV00020B/2040
9781733795920